Statistical Models *in* Earth Sciences

Statistical Models *in* Earth Sciences

Prof. B.K.SAHU
Department of Earth Sciences,
IIT Bombay.

BSP **BS Publications**
A unit of **BSP Books Pvt., Ltd.**
4-4-309/316, Giriraj Lane, Sultan Bazar,
Hyderabad - 500 095
Phone : 040 - 23445605, 23445688

Published by :

BS Publications
A unit of **BSP Books Pvt., Ltd.**

4-4-309/316, Giriraj Lane, Sultan Bazar,
Hyderabad - 500 095
Phone : 040 - 23445605, 23445688
e-mail : info@bspbooks.net

ISBN: 978-93-52300-32-7

Dedicated to

My wife Kanchan ...

Preface

This book represents material that I have been using in my Data Processing course to M.Sc. (Applied Geology) students for the last 23 years. Most of the geological data are not of deterministic type and hence require some probabilistic models due to variations in the samples as well as repeated measurements made on these samples. It is well known that most of the students of earth sciences lack background in mathematics and statistics but have some experience in Differential and Integral calculus. The main purpose of this course was to sharpen their minds for formulating geologically realistic models and to test hypothesis by appropriate statistical methodology. Although geological processes and products are extremely complex often requiring modeling for dependent samples and nonlinearities, it was felt that at the M.Sc. stage it would be pedagogically sound to introduce simpler methods based on independence of samples and linear processes.

We give some introduction to the nature of complexities and processes within earth sciences and how some of these can be modelled through probabilistic and statistical methods using linear processes. These can be best exemplified through the use of axiomatic probability using univariate statistical methodology for scalar random variables and multivariate statistical methods for vector random variables. We also include several real geological examples, which have been solved or partially solved to exemplify applications for solution of specific earth science problems. The aim of these examples is to help the student to conceptually appreciate the problem and realistically formulate a simple mathematical model for its solution. In addition, several exercises are given at the end of each chapter so that the student can independently solve some advanced problems. It is hoped that the book will also be very useful to professional earth scientists in formulating models and using statistical methods to make appropriate decisions in their chosen field.

It is planned to have another book based on my courses at M.Tech. and Ph.D. levels for geology students where "*Time Series and Stochastic Models*" would be emphasized using Linear as well as Nonlinear processes. These advanced models, now published with the title, Time Series Modelling in Earth Sciences, Lisse Netherlands, AA Balkema in 2003, 284 p, do not necessarily assume independence of samples or identical probability distributions for the sample values.

I thank CSIR, New Delhi for financial assistance through award of Emeritus Professorship for the tenure of writing this book. I thank the authorities of IIT Bombay and especially of the Department of Earth Sciences for providing congenial environment and suitable infrastructure so that the book could be completed. I also thank my M.Sc., M.Tech., and Ph.D. students who have contributed to my pursuit of realistic geological modelling for fun and profit. I thank Shri Bhaskar Sahoo, JRF, for computer processing of the manuscript. I shall greatly appreciate if any mistakes are brought to my notice for correction.

IIT Bombay,
Mumbai - 400076
Aug. 2005

Prof. B. K. Sahu
CSIR, Emeritus Scientist (Professor)
(1997-2002) and Emeritus Professor for LIFE,
at IIT Bobmay, Powai, Mumbai, India.

Contents

CHAPTER 1

Introduction

CHAPTER 2

Univariate Distributions

CHAPTER 3

Estimation and Tests of Hypothesis

CHAPTER 4

Linear Statistical Models

CHAPTER 5

Multivariate Analysis

CHAPTER 6

Principal Component and Factor Analysis

CHAPTER 7

Multiple Regression and Cannonical Correlations

CHAPTER 8

Manova, Discrimination and Classification

CHAPTER 9

Some Advanced Statistical Methods

1 Introduction

1.1 Earth and its Evolution

Geology includes static as well as dynamic processes in local, regional and global scales. These could be studied in one, two or three dimensions either evolving or static with respect to time. These processes are often extremely complex to be modelled by simple univariate statistical models but are more realistically multivariate statistical stochastic process models. These advanced stochastic/ time series models allow structural dependency among neighboring sample values in space/time as well as among the different space variables for multivariate situations.

A model is a simplified version of a real complex phenomenon, which can be mathematically defined and therefore predictions from it must be validated through actual measurements (data). Theory of statistics or stochastic processes facilitates the validation of model predictions as well as of model parameters. Depending on the purpose and precision required, it is possible to fit (on a fixed data set) simple to complicated models with fewer to larger number of parameters for a given phenomenon studied. Realistic models should confirm to simple mathematics but satisfying the characteristics of Earth's materials and processes.

Earth is a nearly spherical planet with radius of 6350 km and circumference of 40000 km. It is geologically evolving through the last 4.6 b.y. to have produced a well-differentiated structure: atmosphere, hydrosphere and lithosphere with a thin crust overlying the mantle (lower and upper), core (solid inner and liquid outer cores)). The lithosphere comprising, oceanic crust (6 km thick) and continental crust (40 km thick), overlies the upper mantle with a transition zone having plastic/deformable asthenosphere. Lithospheric plates move over the asthenosphere, to produce diverging, converging and lateral /transform faulted boundaries.

Plate tectonics is in operation since Archean, but the present continental/ oceanic margins are easily reconstructed from Jurassic (200 m.y.). The details of differentiated layered structure of Earth's interior up to present time is given below with depths from the surface of Earth.

	Depth
Lithosphere	0-70 km
Asthenosphere	70-250 km
Transition zone	350-700 km
Lower mantle	700-2900 km
Liquid Iron core	2900-4970 km
Solid Iron core	4970-6370 km
Continental crust	0-40 km
Oceanic crust	0-10 km

Based on these characteristics of Earth, local, regional or global scale models with respect to time and space may be delineated as follows :

Variable	Global/Continental > 10%	Regional 1-10%	Local <1%
Time Range = 0 - 4.6 b.y	>2.27 m.y	0.227-2.27 m.y	0-0.227 m.y
Depth/Elevation (Z) Range = 0 - 12700km	>1270 km	127-1270 km	0-127 km
Surface Horizontal (X) Range = 0 - 40000km	>4000 km	400-4000 km	0-400 km
Surface Horizontal (Y) Range = 0 - 40000km	>4000 km	400-4000 km	0-400 km

Modeling of geological bodies in global/continental scales should be based on spherical projections and/ or spherical harmonics whereas in regional scales it could be approximated using the usual Euclidean rectangular geometry. Modeling of geological bodies in local scales is normally accurate through the use of the usual Euclidean rectangular geometry. However, for some units, precisionisverypoor as we go back in time near to origin of earth and a logarithmic precision scale is used to discriminate among local, regional and global time scales.

In Hadean era (3.8 to 4.0 b.y.) earth had molten silicate surface which cooled to give a solid crust. Atmosphere was dominated by CO_2 with no free molecular oxygen. There was liquid water in hydrosphere with no trace of life. Comets and chondrites brought in organic materials.

In Archean era (3.8 to 2.5 b.y.) life was generated from proteins, nucleic acids, carbohydrates and lipids. Photosynthesis changed the composition of original reducing atmosphere to give free molecular oxygen. Life forms developed as bacteria, eukarye, archaea and prokaryotic life diversified. These made stable and more suitable atmosphere and hydrosphere. In Proterozoic era (2.5 to 0.57 b.y.) prokaryotes gave rise to bacteria and diversified prokaryotes yielded eukaryotic cells (with/without chlorophyll) and later evolved to simple cells. Extinctions are partly internal and partly external but not simple Darwinian theory of natural selection.

In Phanerozoic (0 to 0.6 b.y.) we see a great diversity of plants (eucaryote) and animals because of increased oxygen in atmosphere, surface layer and deeper parts of oceans. Catastrophic extinction occurred at the end of Cretaceous (65 m.y) due to Chixulumb impact cratering near Gulf of Mexico.

1.2 Probability and Statistics

Probability is the foundation for statistics and can be independently used without statistics. In everyday usage, probability expresses a degree of belief one has about an event/statement by a number within zero to unity (Bayesian). However, classical school refers to probability of an event, which can be repeated and interprets it as the limits of the empirical frequency as the number of repetitions tends to infinity. These two methods are complementary depending on situations.

Classical probability was studied with simple events having same probability under repetitions but this were not sufficient to yield probability measure for very complex events. Therefore axiomatic theory was introduced by Kolmogorov, which was more general. Modern probability theory measure relative frequencies which averages the probability when number of repetitions tend to infinity. Probability can also be expressed as betting odds in zero to infinity scale. Here, in this book we interpret probability in terms of classical non-Bayesian method.

Statistics is the science of assigning a probability to an event on the basis of experiments. Often we cannot perform the experiments but are limited to existing population, which can be measured. So, statistics can also be defined as the series of binary data and making inferences with the characteristics of a random mechanism that has generated the data.

A random mechanism giving a real number for each outcome is called random variable. Random variable could be discrete or continuous. A discrete random variable is characterized by the probabilities of the outcomes i.e. probability mass function (p.m.f.) whereas a continuous random variable is characterized by the probability density function (p.d.f.) over the interval. Probability distribution is the cumulative probability and it refers to both discrete and continuous random variables. Therefore, statistics is the science of estimating probability distribution of a random variable on the basis of repeated observation drawn from the same random variable.

1.3 The Statistical Method

The statistical method should include the following steps :

(i) *Problem/purpose :* Units, Target population (Process)
Response and explanatory variables, Population attributes
Problem/purpose : cause, description, prediction.

(ii) *Planning :* Study population/process, its units, variables, attributes.
Selecting response variables and adjusting explanatory variables.
Sampling and measurement processes
Data collection format

(iii) *Data :* Execute the conceptual plan and record any discrepancy/deviations
Data monitoring and internal consistency
Data storage

(iv) *Analysis :* Data summary - numerical and graphical
Modeling: building, parameter fitting, checking cycle
Formal analysis

(v) *Conclusions :* Synthesis : description, graphics
Limitations; potential errors, future improvements

1.4 Population and Sample

Population refers to, any large collection of elements about which we want some information. Population could be finite, countably finite or infinite. Often we cannot measure all the elements of population and it is prudent to select a small finite sample for measurement and study. Since important decisions are made on the basis of samples, it is useful that the samples are characterizing the population and give precise and reliable information about the population.

A sample should be easily obtained and should yield consistent, unbiased, precise (efficient) and sufficient statistics for the population parameters of our interests. Different sampling schemes include:

(i) Simple random sampling (*n/N*) for homogeneous finite population
(ii) Stratified sampling for heterogeneous/stratified population
(iii) Cluster sampling to reduce cost of sampling. We take random sample of groups/clusters and then take sub-sets from the clusters selected.
(iv) Systematic sampling for the population having fixed order
(v) Double sampling. Initial sample size yields an estimate of lower precision. These statistics are then used to find the number of additional items to be measured in order to finally yield an estimate with the required confidence limits to final estimate.
(vi) Sequential sampling of fixed size for quality control.

Sample size should be sufficiently large to yield precise statistics but should be sufficiently small with respect to population size, so that the estimates are unbiased population parameters. For example, sample size (*n*) should be more than twenty for obtaining precise statistics but should be less than 10% *N* (*N* = population size) to yield unbiased population parameters. Otherwise, finite population corrections to sample statistics are to be made.

Sampling : Concepts and Methods

Sampling is undertaken to obtain information about a population without examining every item. Population is large collection of elements such as people, farms, coin-tosses and incomes etc., about which information is desired. Sampling theory is concerned with ways of estimating and improving the precision of the information obtained from the sample.

For statastical theory of estimation hypothesis testing and decision, we require random sample and use theory of probability. A sample of given size is called random if every possible sample of this size in the population has a calculable probability of being selected although this probability may vary from item to item. In case of simple random sampling, the probability of each item in sample of size *N* has equal probability. Theoretical statistics is based on the concept of simple random sampling but in practice some modifications are adopted due to costs, convenience and requirement of greater precision.

If prior information is available, it will be useful to gain precision through stratified sampling where the population is divided into homogeneous groups and a simple random sampling adopted for each group (strata). Such a method would be useful for sampling a stratigraphic section, borehole core etc., over which a homogeneous strata structure would be superposed, so that variance of each strata is representative and overall variance is deduced.

Sample surveys are conducted in such a way as to reduce overall cost through cluster sampling. Usually sub-samples are collected from the clusters selected by initial random sampling. Although this may not be efficient as regards precision of estimates, it gives good results when the population is highly diversified (opposite of stratified population).

Another method to reduce costs is by systematic sampling (house in blocks) but some care should be taken so that sample is representative and not unduly biased.

In opinion surveys, we often use quota system over sex, age group, income-group etc. This method may introduce an unknown bias for estimates. In purposive sampling a small size is usually selected, which is not random sample. But the sampler should have good judgement and expert knowledge in such cases, otherwise results would be unreliable.

Often we are forced to take sample that is available, which need not be random sample. For example, we are forced to sample available rocks and ores and cannot sample material, which has been eroded away. Nevertheless, a probability sample should be obtained whenever possible. Double sampling is often used to reduce the cost and obtain required precision in the estimate. Suppose we get x_1 and σ_1^2 based on initial N_1 samples and require x_1 to lie within $\mu \pm d$ at 0.05% level. Then $N_2 > N_1$ required to achieve 0.05% level confidence is given by

$$N_2 = \frac{\sigma_1^2 \, t^2_{N_1 - 1, .05}}{N_1 - 1}$$

so that additional of number of samples ($= N_2 - N_1$) are to be collected. In remote areas, however, double sampling would not be feasible and adequate sample size should be collected in first stage itself.

1.5 Model Types

Models can be classified into following types :

(i) Deterministic, probabilistic models.

(ii) Linear or nonlinear models, where the functional relations could be either linear or nonlinear.

(iii) Discrete, continuous or combination random variable models.

(iv) Univariate or multivariate statistical models, where the number of random variable is one or more than one.

(v) Models could be static or dynamic (time-varying). For example, stochastic process and time series models are dynamic type.

(vi) Models could be spatial in one, two or three dimensions with or without time varying effects. For example, atmospheric pollution is 3D time varying phenomenon whereas ore mineralisation is a 3D static phenomenon.

Most of these model types could be operated independently so that in all we have $2^3 \times 3^2 = 72$ possible gross model types in earth sciences. If samples are observed at wide spacings in time and/or spatial domain, these tend to be independent and yield statistical models (univariate or multivariate) whereas close spaced sampling in time and/or spatial domain gives time series models (univariate or multivariate).

Models should be mathematically simple enough to be at least approximately solvable but should include all important characteristics of the complex phenomenon to be realistically described/ predicted. Therefore, with more data or for different purposes, the models for a given phenomenon could be different.

1.6 Characteristics of Earth Sciences Data

Data could have different scales of information and can be classified as follows in increasing trend of information.

(i) *Nominal :* different from each other but no ordering

(ii) *Ordinal :* different from each other but in a given order

(iii) *Interval :* intervals have same information independent of the level of the data

(iv) *Ratio :* scale has fixed zero

Nominal and ordinal scale data are discrete variables whereas intervals and ratio scales are continuous variables. In order to retain more information, measurements should preferably be expressed in ratio scale for continuous data and ordinal scale for discrete data. Logarithmic transformation of a nonzero positive real variable makes the transformed data less informative.

Geological data analysis includes several stages in a complete cycle such as data elaboration, data collection, data management, data exploration, data synthesis, data communication, decision making and finally communication of these decisions. Within geological data exploration, are included 1*D* exploration, 2*D* exploration, *N-D* exploration, analogy, data clustering etc.

Most geological phenomena/processes are extremely complex with strong interaction among the various component elements of the system. Models reduce the phenomena into simple subsystem with strong interaction within them but weak interactions (no interaction) among subsystems. This considerably simplifies mathematical / statistical / stochastic modeling of original complex system.

1.7 Axiomatic Probability

It is based on the concepts of sample space, event and probability measure for the event. Sample space is the set of all possible outcomes of an experiment. Event is defined as a subset of sample space inclusive of the total sample space. Simple event cannot be formed by union of other events. Composite events are not simple events.

Probability space is represented by the triplet (Ω, A, P), where Ω is the collection of points associated with the outcomes of random phenomena, A is the sigma field of subsets A of Ω which are closed under countably infinite unions, intersections and complementation. Hence, probability measure P of an event A lies between $0 \leq P(A) \leq 1$, i.e. it is a real valued function having domain A with the following parameters :

(i) $P(\Omega) = 1$

(ii) $P(A) \geq 0$ for all $A \varepsilon \Omega$

(iii) If A_n, $n = 1, 2, 3 \ldots$ are mutually disjoint sets in Ω then

$$P\left(\cup_{n=1} A_n\right) = \Sigma_{n=1} P(A_n)$$

$$P(A \cup B) = P(A) + P(B) - P(A \cap B)$$

However, if A and B are mutually exclusive events (independent) then,

$$P(A \cup B) = P(A) + P(B)$$

since $P(A \cap B) = P(\phi) = 0$, where ϕ is the null set.

Also $P(A^c) = 1 - P(A)$ and in particular $P(\phi) = 1 - P(\Omega) = 0$

De Morgan's law states $(\cup_n A_n)^c = (\cap_n A_n^c)$ and $(\cap_n A_n)^c = (\cup_n A_n^c)$.

Hence,

$$P(\cup_n A_n) = 1 - P(\cap_n A_n^c) \text{ (Fig. 1.1)}$$

In some cases it is easier to compute

$$P(\cap_n A_n^c) \text{ rather than } P(\cup_n A_n).$$

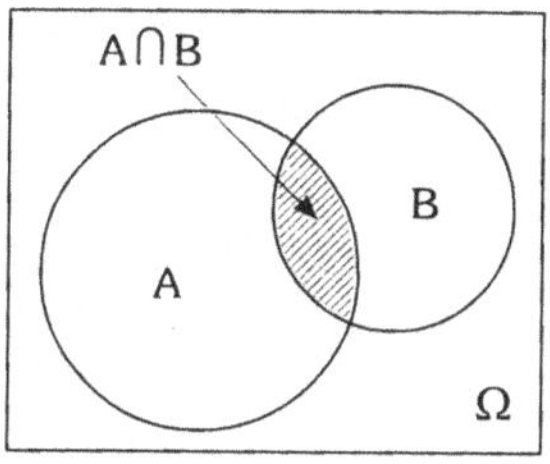

Fig. 1.1 Universe (Ω), sets A and B intersection $A \cap B$

Conditional Probability

We assume $P(B) > 0$. Then the four axioms are

(i) $P(A|B)$ ^{3} 0

(ii) $P(A|B) = 1$ for $A \supset B$

(iii) If $\{A_i \cap B\}$, $i = 1, 2, 3$ are mutually exclusive, then
$P(A_1 \cup A_2 \cup \ldots .. / B) = P(A_1|B) + P(A_2|B) + \ldots$

(iv) If $B|H$ and $B|G$ and $P(G) \neq 0$, then $P(H|B)/P(G|B) = P(H)/P(G)$

The four axiomatic rules for conditional probability reduces to

$$P(A|B) = P(A \cap B)/P(B) \text{ provided } P(B) \neq 0.$$

If events A and B are mutually independent events then $P(A \cap B) = P(A).P(B)$ and this can be extended to infinite sets of mutually independent events. Pairwise independence is not sufficient for mutual independence of all the sets.

Some Geologic Applications

(i) *Discrete probability :*

(a) Point counts modal analyses of minerals are having binomial distribution.

(b) Heavy mineral counts have Poisson distribution.

(ii) *Continuous probability :* Size distribution of grains in river bed sediments

(a) unimodal positive skewness => log transformed gaussian.

(b) bimodal/polymodal due to deposition from bed, saltation and suspended loads. It is dissected into 2 or more gaussian models each having its characteristic mean and variance and its components frequency weights summing to 1.0

(iii) *Mixed probability :* Thin section size analysis

(a) A few grains(n_1)are lying in thin section plane having discrete probability (α / (1 + α)) of being intersected by thin section plane with a probability 1.0. For these grains we do not need bias correction. (Krumbein-Hagermann model).

(b) Most grains$(N-n_1)$ lying out side the thin section plane with probability of intersection $1/(\alpha)$. The number of grains intersected in thin section is $(N-n_1)/(1+\alpha)$ and since thin section plane do not pass through the grain centre, the observed sizes of sectioned grains are smaller in size and need to be bias corrected (Wicksell's model).

1.8 Examples

Example 1

A pair of fair dice is rolled once. What is the probability that sum is 7 ?

Answer : Sum 7 can occur as (1, 6), (2, 5), (3, 4), (4, 3), (5, 2), (6, 1).

The total sample space is $6^2 = 36. P(7) = 6/36 = 1/6$.

Example 2

If 5 cards are drawn, what is the probability of 3 aces ?

Answer : P (3 Aces) = $(4\ C_3)\ /\ (^{52}C_5)$

Sample space is $^{52}C_5 = 94/54145 \approx 0.001736$.

Alternatively, we can use conditional probability and independence.

Consider first 3 cards are aces (A_i) and last two non-aces (N_i).

Then,
$$P(A_1 \cap A_2 \cap A_3 \cap N_4 \cap N_5)$$
$$= P(A_1)\,P(A_2 \mid A_1)\,P(A_3 \mid A_1 \cap A_2)\,P(N_4 \mid A_1 \cap A_2 \cap A_3)$$
$$P(N_5 \mid A_1 \cap A_2 \cap A_3 \cap N_4)$$
$$= 4/52\ .\ 3/51\ .\ 2/50\ .\ 48/49\ .\ 47/48$$

But 3 aces in 5 cards can occur in $^5C_3$ ways.

Hence, the probability is

$$^5C_3\ 4/52\ .\ 3/51\ .2/50\ .\ 48/49\ .\ 47/48 \approx 0.001736,$$

which is same as that obtained by first method.

Example 3 *(Pairwise independence is not same as mutual independence)*

Let A be event of H in first toss, B the event of H in second toss and C the event that both tosses of coin yield either heads (H) or tails (T).

Then, events A, B, C are pairwise independent.

$$P(A \cap B \cap C) = P(A \cap B) = 1/4$$

But $P(A)\ .\ P(B)\ .\ P(C) = 1/8$ (not equal to 1/4).

Hence, events A, B, C are not mutually independent.

Example 4 *(Bayes' Theorem)*

Let diagnostic test be 95% accurate both for patients and non-patients. Assuming the population affected 0.005, what is the probability that a particular person is affected given that the test is positive for him.

Answer : Let C be the event that he is affected and T be the event that test is positive. Then according to *Bayes' Theorem.*

$$P(C \mid T) = P(T \mid C)\,P(C) \,/\, [P(T \mid C)\,P(C) + P(T \mid C)\,P(C)]$$

$$= \frac{0.95 \times 0.005}{0.95 \times 0.005 + 0.005 \times .995}$$

$$\approx 0.087$$

Example 5

Find the probability that no two students have same birthday in a class having 25 students ?

Answer : $p = (1-1/365)(1-2/365) \;.......\; (1-24/365)$

Writing $\log_e (1-x) \approx -x$, we get

$\log_e p = -(1 + 2 + 3 + \;.....\; + 24)/365$

$= -0.823$ or $p = 0.44$

Example 6

How many times should a card be drawn and replaced, so that there is 90% chance ofseeing the ace of spades at least once?

Answer : The chance of not seeing ace of spades in n draws is $(51/52)^n = 0.1$.

So, $n\,(\log 52 - \log 51) = \log 10 = 1.$

We get, $[n] = 1/0.0084 \approx 119.$

We take n as an integer value which is a little higher than the exact inverse.

Example 7

If 10,000 tickets are sold in a lottery and there are prizes of Rs.1000(one); Rs.50(twenty). What is the worth (expectation) for one ticket ?

Answer : Expectation of a ticket

$$= \frac{\text{Rs}1000}{10,000} + \frac{\text{Rs}50 \times 20}{10,000} = \text{Rs. } 0.20.$$

So fair price of a ticket should be Rs. 0.20 and not higher.

Example 8

A needle of length l is dropped on a ruled board having parallel lines a ($\geq l$) apart. Find the probability that the needle crosses a line.

Answer : The needle will cross the nearest line if distance of its centre is less that $\left[\frac{1}{2}\ l\ sin\theta\right]$, where θ is angle of needle with the line.

$$p = \int_0^{\pi} \int_0^{(1/2) l \sin\theta} dy\, d\theta = \frac{1}{\pi a} \int_0^{1} sin\theta\ d\theta = \frac{2l}{\pi a}$$

Example 9

In sieving sedimentary particles through square mesh, what is the probability of passing through the mesh with aperture L ?

Answer : A needle like grain can pass diagonally (45^0 to wires) even if $l = \sqrt{2}\ L$, hence position of grain and its orientation (θ) to a wire mesh are not independent. The grain passes the square mesh simultaneously at angle θ to one set and 90 $-\theta$ to other set of wires, orthogonal to first set of wires.

$$p = \int \left(1 - \frac{p(a,b,c;\theta)}{L}\right)\left(1 - \frac{p(a,b,c;90-\theta)}{L} d\theta\right)$$

p ($<L$) is the projection of grain parallel to one set of wires and it depends on sizes a, b, c and orientation θ.

1.9 Case Studies

Thin Section Analysis

A thin section of rock/ore is cut for model and size analysis in lower $2D$ or $1D$ space from which inference on these properties are made in $3D$ space using statistical and stereological procedures. Since the section plane does not necessarily pass through grain-centre of minerals, the measured sizes and size-distributions are highly biased towards smaller sizes which require probabilistic corrections depending on model employed. In addition, larger grains have higher probability of being cut and measured in lower space.

A few grains whose centres lie on the section do not require any bias corrections. These grains can be modelled as discrete probability (Poisson Distribution) in the concerned space. However, the number/proportion of such grains are unknown and has to be estimated.

Major proportion of grains have centres outside the section plane but are included conditional on their perpendicular distance being less than their radius. The random distribution from plane is a continuous variable ranging in semi-closed interval [0, R); R being the maximum radius of grains in the sample. The number/proportion of grains cut by section can be estimated by integrating the conditional probability over distance (assumed to have uniform distribution).

Realistically, the grains cut by section plane are sum of both probabilities i.e. mixture probability (α for discrete part and (1-α) for continuous part). Hence, true size distribution in higher space should be obtained by weighted average of discrete and continuous probability. The Hagerman-Krumbein model assume all grains have equal probability irrespective of their sizes. This is equivalent to discrete probability (α) and probability size corrections are ($j \geq 0$),

$$n_{x,j} \left\{ \int_0^{\pi/2} sin^{j+1} \theta \, d\theta \right\}^{-1} . \ (H\text{–}K \text{ model})$$

where $n_{r,j}$ and $n_{x,j}$ are the jth order moments of r (sphere radius) and x (circle radius) respectively.

The Wicksell model assumes the probability of inclusions in the section plane is directly proportional to sphere radius (r) and probability corrections are ($j \geq 0$),

$$n_{r.j} \left\{ \int_0^{\pi/2} sin^{j} \, \theta d\theta \right\}^{-1} . \ n_{(j-1)} \ \pi/2 \, . \, n_{x-1} \text{ (Wicksell model)}$$

where $n_{x,-1}$ is harmonic mean of x.

Sahu (1974a) developed the realistic mixed probability model and the moments of r are given by the weighted average,

$$n_{r,j} \quad = (1/\alpha + 1) \ [\alpha \ H\text{-}K \text{ model} + W \text{ model}] \text{ for } j > 1.$$

The first form number frequency moments are

$$n_{r,1} \quad = (1/\alpha + 1) \ [\alpha \ (4/\pi) \ n_{x,1} \ + \ (\pi/2) \ n_{x,-1}]$$

$$n_{r,2} \quad = (1/\alpha + 1) \ [\alpha \ (3/2) \ n_{x,2} + (4/\pi) \ n_{x,1} \ . \ (\pi/2) \ n_{x,-1}]$$

$$n_{r,3} \quad = (1/\alpha + 1) \ [\alpha \ (16/3\pi) \ n_{r,3} + (3/2) \ n_{x,2} \ . \ (\pi/2) \ n_{x,-1}]$$

$$n_{r,4} \quad = (1/\alpha + 1) \ [\alpha \ (15/8) \ n_{r,4} + (16/ \ 3\pi) n_{r,3} \ . \ (\pi/2) \ n_{x,-1}]$$

These number frequency moments can be converted to volume, weight (ρ being constant) frequency moments as, $w_{r,j} = n_{r,(3+j)}/n_{r,3}$ $(j \geq 0)$.

If size of grains is measured as half-intercepts (t) along random lines on section plane, we can obtain moments of circle radii from moments of t by using the above equations. But since $n_{x,-1}$ is not available we cannot compute $n_{r,j}$ from knowledge of $n_{t,j}$ directly. This is because knowledge of t is conditional on the fact that the section has cut the $3D$ grains to provide with x data. Hence, the stereological formula converting moments of t directly to moments of r are not correct. Sahu (1974b) solved this problem by computing $n_{x,-1}$ (mm) assuming the $n_{x,j}$ are distributed as approximately phi-normal curve ($phi = -log_2$ (*diameter in mm*)), so that Edgeworth's expansion is applicable and reasonably accurate. We get,

$$n_{x,-1}(mm) = e^{-\mu(\ln 2) - 1/2(\ln 2)^2\sigma^2} [1 - 1/3!(\mu_3/\sigma^3)(\ln 2)^3\sigma^3 + 1/4!\{(\mu_4/\sigma^4) - 3\}\ln 2)^4\sigma^4 + 10/6!\ (\mu_3/\sigma^3)^2(\ln 2)^6\sigma^6]^{-1}$$

where μ = phi mean,

σ^2 = phi variance,

μ_3/σ^3 = phi skewness,

and $\{(\mu_4/\sigma^4) - 3\}$ = phi-kurtosis which can be obtained for $n_{x,j}$ $(j \geq 1)$.

Conversion of equations from arithmetic to phi-moments are given by Sahu (1965) (equ^{n}. 22,21,23,24; note that equ^{n}s 23, 24 should be multiplied by -1 in their RHS) as follows :

$$\sigma^2 = (\log n_{x,2} - 2 \log n_{x,1})/(\log 2)(\ln 2)$$

$$\mu = (1/2 \log n_{x,2} - 2 \log n_{x,1})/(\log 2)$$

$$\text{Phi skewness, } \mu_3/\sigma^3 = -\log_2 n_{x,3} - 3\mu + 1/2\ (\ln 2)\ \sigma^2$$

$$\text{Phi kurtosis, } (\mu_4/\sigma^4) = -\log_2 n_{x,4} - 4\mu + 8\ (\ln 2)\ \sigma^2 + 3.0$$

Thin section methodology is eminently suitable for study of granular porous media such as sandstones, siltstones, grits, limestones, dolomites etc. where prime interest is prediction of $3D$ porosity and permeability tensor. If three orthogonal sections are cut then the porosity and permeability measured in each section can be transformed to $3D$ measure by suitable probabilistic correction model.

If ϕ_1, ϕ_2 and ϕ_3 are porosity measured in each of the orthogonal sections and it is known *stereological theory*, these are equivalent to $3D$ porosities. Therefore, a scalar (isotropic) porosity measure in $3D$ would be $\phi = (\phi_1 \phi_2 \phi_3)^{1/3}$ where ϕ is the geometric mean of the measured (independent) three porosities.

If individual grains vary in their size as well as sphericities, Sahu (1974c) has shown that specific surface (*Ss*) can be computed (by superposing a square grid) as :

$$Ss = 2m\ Ig\ Nt\ (1 - \phi)/Lt\ Ng$$

where, m = linear magnification

Ng = mesh points lying within grains

Nt = total number of mesh points

Ig = number of intersection between lines and perimeter of grains

Lt = total length of grid lines.

The specific surface is related to intrinsic permeability (k) as

$$k = CS_s^{-2}$$

where C is constant.

Therefore, we can predict permeability k perpendicular to thin section plane once C is estimated. The permeabilities so obtained can be arranged as a permeability tensor (matrix).

Optimal Sample Size in Lower space

The problem was solved empirically by many workers but their solutions did not have any theoretical basis. Statistically 20 counts and 300 counts in $3D$ are large and sufficient sample sizes, respectively. But sizes are measured in lower space say $2D$ or $1D$ by thin section procedure and we should have a concept of equivalent precision in higher space(*s*) after the probabilistic corrections which were applied. Sahu (1977) attacked this problem by analyzing arithmetic moment $n_{r,1}$ in higher space as (Sahu model) follows :

$$n_{r,1} = [1/(\alpha + 1)]\ [\alpha\ (4/p)\ n_{x,1} + (p/2)\ n_{x,-1}]$$

where $\alpha = R/r$ is a constant for a given section (sample). α can be estimated either by direct measurements in higher space or by iteration of initial $\alpha_0 = x_{max}/(4/\pi)\ n_{x,1}$. Analysis of $n_{x,1}$, $n_{x,-1}$ in sub samples circle radii counts yielded a linear regression of log(N) vs σ_N^2 with 0.01% significance as :

$$\log N = 2.0964 - 0.8876\ \sigma_N^2 \quad \text{(for } x \text{ to } r \text{ data)}$$

Similarly, for t to x data with subsamples 55,110 ... 660, we get

$$\log N = 2.8704 - 0.4507\ \sigma_N^2 \quad \text{(for } t \text{ to } x \text{ data)}$$

Using these linear equations we can predict log N (hence the optimal sample size N) required to give the some precision of corrected arithmetic mean in higher space. The result gave that for $2D$ to $3D$ space we need 13 times more circle diameters. Similarly for $1D$ to $2D$ space, we also need the same 13 times more intercepts to reach same precision. Therefore, we conclude that for space reduction of 1 inch, we require 13 times more measurements in lower space.

Hence, for prediction in $3D$ space from half intercepts ($2D$ space), we require $13 \times 13 = 169$ times more measurements than sphere size data. Thus, 100 sphere are equivalent in precision to 1300 circle diameters and 16900 half-intercepts.

Sieving Particles through Square Mesh

Sieving is described as '*Cinderalla*' of sizing process as it performs most of the hard work yet is greatly neglected as regards research regarding accuracy and precision. Natural sedimentary particles may be classified as either ellipsoidal or prismatic with rounded corners and their probability of passage through square mesh depends largely on size, sphericity and roundness. Hence, equivalent sphere diameter may be defined based on probability of passage through square mesh (Sahu 1975,1976, 1980).

Mechanics of Ro-Tap sieving has two cyclic motions:(1) Circular (Horizontal) motion for large time T_1 and (2) Tapping (Vertical) motion for short time T_2. During horizontal motion the grain presents in its larger cross-section (A-B) for passage whereas during vertical motion it presents its shorter cross-section (B-C) for passage. The average probability of passage for the $3D$ grain is the weighted time average of the two probabilities, of which BC probability of passage is dominant. In the literature, particle size is incorrectly assigned to aperture L of square mesh through which it passes.

So, we have $\bar{p}(ABC) = [t_1\, p(BC) + t_2\, p(AB)] \,/\, (t_1 + t_2)$.

We have defined size of AB section as $(AB)^{1/2}$, sphericity as (B/A) and roundness (ρ). Similarly, for BC section size is $(BC)^{1/2}$, sphericity as (C/B) and roundness (ρ). For convenience, we compute probability of passage for elliptical outlines for which roundness ρ (= sphericity B/A or C/B) does not affect passage probability.

If long axis of ellipse is oriented at angle θ to one set of wire mesh (L apart), the projection diameter $P(\theta)$ perpendicular to that set is given by,

$$P\,(\theta)/A = [1 - \{1 - (B/A)^2\}\cos^2\theta]^{1/2} \qquad .....(1.1)$$

for section. For BC section replace A by B in LHS and B/A by C/B in RHS. The probability of passage of ellipse through square mesh of aperture (L) is given by,

$$P\,(AB;\,\theta) = [1 - P\,(\theta)\,/\,L]\,[1 - P\,(90 - \theta)\,/\,L] \qquad .....(1.2)$$

Since the ellipse has to pass through two perpendicular sets of wires L apart simultaneously (i.e. if long axis is at angle θ to one set, it is at angle $(90 - \theta)$ to its perpendicular set). However, in the literature, the probability of passage through square mesh has been erroneously independent for each set of wires.

For passage of *BC* section, we replace *AB* by *BC* in LHS and substitute $P(\theta)/L$ and $P\,(90 - \theta)\,L$ appropriately, as given by eqn. 1. The nuisance parameter θ can be eliminated by numerically integrating eqn. 2 numerically from 0° to 90° at small angular intervals ($d\theta$ say 5^0).

Table 1.1 gives the results for $\bar{p}(AB)$ or $\bar{p}(BC)$ for various sizes ($\sqrt{AB}$ or $\sqrt{BC}$) and sphericities (*B*/*A* or *C*/*B*) for ellipses. Sahu (1976) includes Table I and Table II for projection diameter $P(\theta)$ of ellipse and prismatic outlines with different roundness values (ρ) and different θ values.

Table 1.1 *Values of* $\bar{p}$ *(AB) or* $\bar{p}$ *(BC) for different size and sphericities*

Avg. Sph	0.1	0.2	0.3	0.4	0.5	0.6	0.7	0.8	0.84	0.9	1.0
.025	.54	.00	.00	.00	.00	.00	.00	.00	.00	.00	0
.125	.88	.61	.27	.02	.00	.00	.00	.00	.00	.00	0
.225	.93	.75	.51	.26	.04	.00	.00	.00	.00	.00	0
.325	.95	.81	.62	.41	.20	.03	.00	.00	.00	.00	0
.425	.95	.83	.67	.48	.30	.13	.02	.00	.00	.00	0
.525	.96	.85	.70	.53	.36	.20	.07	.01	.00	.00	0
.625	.96	.86	.72	.56	.40	.24	.12	.02	.01	.00	0
.725	.96	.87	.73	.58	.42	.27	.15	.05	.02	.00	0
.825	.96	.87	.74	.59	.43	.29	.16	.07	.04	.01	0
.925	.96	.87	.74	.59	.44	.29	.17	.08	.05	.02	0
1.000	.96	.87	.74	.59	.44	.29	.17	.08	.05	.02	0

Equivalent sphere diameter *D* having the same probability of passage as the particle (*ABC*) is given by,

$$D = L(1 - \bar{p}(ABC)^{1/2}) \quad \text{.....(1.3)}$$

Table 1. 2(a) Values of *P*/*A* for different values of θ and *B*/*A* or *C*/*B* for ellipses

$$P/A = [1\text{-}\{1\text{-}(B/A)^2\}\cos^2\theta]^{1/2}$$

θ in degrees B/A or C/B	2.5	7.5	12.5	17.5	22.5	27.5	32.5	37.5	42.5	47.5
.025	.05	.13	.22	.30	.38	.46	.54	.61	.68	.74
.125	.13	.18	.25	.37	.40	.47	.55	.62	.68	.74
.225	.23	.26	.31	.32	.44	.50	.57	.63	.70	.75
.325	.32	.35	.38	.43	.49	.54	.60	.66	.72	.77
.425	.43	.44	.46	.50	.55	.60	.65	.70	.74	.79
.525	.53	.54	.56	.58	.62	.66	.70	.74	.78	.82
.625	.63	.63	.65	.67	.69	.72	.75	.79	.82	.84
.725	.73	.73	.74	.75	.77	.79	.81	.84	.86	.89
.825	.83	.83	.83	.84	.85	.86	.88	.89	.91	.93
.925	.92	.93	.93	.93	.94	.94	.95	.95	.96	.97

Table 1.2(b) Values of factor $(B/A)\cos\theta + \sin\theta$ and $(\cos\theta + \sin\theta - 1)$ for P/A or P/B computation needed in case of rectangular grains. $P/A = B/A\cos\theta + \sin\theta - \rho(B/A)(\cos\theta + \sin\theta - 1)$

θ in degrees / B/A or C/B	2.5	7.5	12.5	17.5	22.5	27.5	32.5	37.5	42.5	47.5
B/A cos θ + Sin θ values										
.025	.069	.153	.241	.325	.406	.484	.558	.629	.694	.754
.125	.169	.254	.339	.420	.498	.573	.643	.708	.768	.822
.225	.268	.354	.436	.515	.591	.661	.727	.787	.892	.889
.325	.368	.453	.534	.611	.683	.750	.811	.867	.915	.957
.425	.468	.552	.631	.706	.775	.839	.896	.946	.989	1.024
.525	.568	.651	.729	.801	.868	.927	.980	1.025	1.063	1.092
.625	.668	.758	.827	.897	.960	1.016	1.064	1.105	1.136	1.160
.725	.768	.849	.924	.992	1.053	1.105	1.149	1.184	1.210	1.227
.825	.868	.949	1.022	1.088	1.145	1.145	1.233	1.263	1.284	1.295
.925	.968	1.048	1.120	1.183	1.282	1.237	1.317	1.34	1.358	1.365
(cos θ + sin θ − 1) values										
	.043	.122	.193	.254	.307	.349	.381	.402	.413	.413

Closure and Normalization of Petrographical and Geochemical Data

Petrographical and geochemical (Mineral/Chemical) proportion (c_i, $0 < c_i < 1$) are not continuous variables. Both are discrete having closure constraints (fixed totals) which includes undesirable negative associations (correlations) among its constituents in rocks and ores. Petrographic modal analysis is usually made by point counts, which have binomial distributions. Similarly, major oxides, major and trace elemental data are binomial (Poisson if proportion is very low). It is well known that binomial proportion is not homogeneous but heterogeneous with variance $[p(1-p)]$ depending on proportion p. These three problems can be simultaneously solved (Sahu, 1995) using $\log[c_i/(1-c_i)]$ transformation for the ith component ($i = 1, 2, 3\ldots$). Such a prior transformation converts it to a multivariate normal distribution with univariate normal marginals, [$N(0)$; var. = 2.567; std.dev. = 1.668], so that parameter estimation and hypothesis testing etc. can be accomplished as the spurious negative correlations do no longer exist. A measure- theoretic proof for normalization using $\log[c_i/(1-c_i)]$ transform is given in Le Cam(1986) and Le Cam and Yang (1990). It should be noted that if $c_i \to 0$ (trace components) then $\log[c_i/(1-c_i)]$ tends to $\log c_i$ (i.e. log normal) as known and desired. Additional advantage of $\log[c_i/(1-c_i)]$ transformation can be listed as follows :

(i) Normalization is made by knowledge of one component c_i only, and knowledge of other components is not required (independence) as was the case in geological literature.

(ii) Since rocks/ores are generally multi-component, we should analyze the data through proper multivariate normal distribution (MND) having univariate normal marginals, principal components and conditional distribution (Regression).

(iii) In the petrogenesis, environmental pollution studies and ore genesis true correlation and partial correlation are geologically meaningful and this is possible through $\log[c_i/(1 - c_i)]$ pre-transform of data sets and using standard MND theory.

1.10 Exercises

1. If $P(A) = 1/3$ and $P(B) = 3/4$, where A and B are known to be independent events, find $P(A \cup B)$.

2. Two fair dice are rolled simultaneously. Let A be the event the 'sum is 8' and B the event the two numbers are same. Find $P(A)$, $P(B)$, $P(A \cap B)$, $P(A \cup B)$, $P(A \mid B)$, $P(B \mid A)$.

3. The host and hostess sit at opposite ends of a long dinner table. In how many ways $2n$ guests can be sit (n on one side) such that two important guests do not sit together ?

4. A man writes n letters and addresses on n envelopes. If letters are placed into envelopes at random, what is the probability that at least one letter is in the correct envelope ? (Ans. for n large it is $1-e^{-1}$).

5. A bridge player and his partner have nine spades. What are the probabilities that the remaining four spades are split between opponents 4-0, 3-1, 2-2 ?

6. Balls numbered 1 to 10 are in an urn. These are drawn one at a time and placed in a straight row of holes numbered 1 to 10. If each ball drawn is placed at its proper place in straight row holes, what is the probability that there will not be an empty hole between two filled ones at anytime of drawing ?

7. A coin is tossed until a head appears and the number of tails obtained is recorded. What is the probability of getting x tails before the first head ? What is the expectation of X ?

8. A diamond of value V is broken into two pieces. If the value of diamond varies as the square of its weight, what is the expected value of the broken diamond ?

9. A stick of length a is broken into three pieces, what is the probability that pieces form a triangle ?

References and Further Reading

1. Le Cam, L., 1986, Asymptotic methods in statistical decision theory, NewYork, Springer, 742p
2. Le Cam, L. and Yang, L., 1990, Asymptotics in statistics, New York, Springer, 180p
3. Sahu, B.K., 1974a, Probabilistic solutions to thin section size measures, Indian Journal of Earthsciences, 1(1) : 22-36
4. Sahu, B.K., 1974b, Probabilistic solutions to thin section size analysis expressing size as intercept and frequency as number or measure. Indian Journal of Earthsciences 1(2) : 168-176
5. Sahu, B.K., 1974c, Determination of average grain sphericity in granular porous media. Journal of Sedimentary Petrology, 44 : 578-582.
6. Sahu, B.K.,1975, Size analysis by sieving technique : Proc. Sympo. on Sediments, Sedimentation and Sedimentary Environment, Delhi University. 1-7
7. Sahu, B.K., 1976, Mathematical theory of counting two-dimensional grains by line and ribbon methods, Sedimentary Geology, 16 : 177-192.
8. Sahu, B.K., 1977, Optimal sample size for thin section size studies, Sed.Geology,19 : 185-196
9. Sahu, B.K., 1980, Mathematical principles of size determination by sieving technique, Proc. European Particle Technology Seminar, Vol.B, Amsterdam, p 805-816
10. Sahu, B.K., 1995, Statistical inference for geochemical and petrographic data, Proc. Recent Research in Geology in Western India, MS University, Baroda, p59-64

2 Univariate Distributions

2.1 Introduction

Once N data are collected from the random variable X in a sample, it is prudent to summarize these large data by a few summary statistics representing the essential features of the sample distribution. If sample size N is small then grouping into small classes is unnecessary but if N is large (say > 100) then grouping could smoothen the distribution and reduce computations.The optimal number of classes (groups) for unimodal distributions should be more than 6 to 7 but should not exceed 20. An empirical formula is $c = 1 + 3.322 \log (N)$, which gives $c = 8$ for $N = 100$, $c = 11$ for $N = 1000$ and $c = 13$ for $N = 10000$.

Graphical representations of frequency, cumulative frequency can be easily plotted with c classes from which median (50 percentile) and other fractiles can be obtained. These give graphical statistics for mean, mode, dispersion, skewness (asymmetry) and kurtosis (peakedness), which describe a unimodal distribution.

Moment computations can be done for N data. N data grouped into k classes for discrete or continuous random variable X as,

$$m'_r = N^{-1} \Sigma_{i=1}^{k} f_i x_i \, \mu'_r = \int x^r f(x) dx$$

where r represents the rth order of moment and $m' = 1$, when $r = 1$ we get the arithmatic average(mean) and $r = 2$ gives the dispersion when mean is at origin. Thus, r^{th} moment about the mean is given by

$$m_r = N^{-1}\Sigma_{i=1}^{k} f_i\left(x_i - m_i'\right) = \int\left(x - m_1'\right) f(x)dx$$

where m_1' is the mean of distribution.

Hence, $m_2 = (m_2' - m_1')^2$ and is called *variance.*

Skewness is computed as $m_3/(m_2)^{3/2}$ and *kurtosis* as $m_4/(m_2)^2$. These are k *statistics* and are better than m *statistics* to represent population parameters. Symbol E is *expectation,* which is the sum or integral of a given function of a random variable weighted over the frequency distribution.

Generating functions generate moments and other character of a population. Moment generating function (m.g.f) $M(h)$ is defined by

$$M(h) = E(e^{hX}) = \sum_j e^{hXj}\, p(Xj)$$

$$= \sum_j (1 + hXj + h^2/2\,!\,(Xj)^2 + ...)_\infty\, p(Xj)$$

$$= 1 + h\,\mu_1' + h^2/2\,!\,(\mu_2') + ... = \sum_{r=0} h^r/r\,!\,(\mu_r')$$

or, $$M(h) = \int_{-\infty}^{\infty}\left(e^{hx}\right) f(x)\, dx$$

Cumulants are logarithms to base e of the mgf

$$K(h) = \ln M(h) = k_1 h + k_2 h^2/2\,! + k_3 h^3/3\,! + ... = \Sigma_{r=1}^{\infty} h^r/r\,!\,(k_r)$$

$k_1 = \mu_1'$ (population mean) and $k_2 = \mu_2$ and $k_3 = \mu_3$ but higher cumulants differ from moments by certain fixed function of earlier moments.

If population mean is taken as $\mu_1' = 0$, and then all $m_r' = \mu_r$.

We have $$M(h) = 1 + \sum_{r=2} h^r/2!\,(\mu_r)$$

$$M(h)/dh = \mu_2 h + (\mu_3)\, h^2/2! + (\mu_4)\, h^3/3! + ...$$

and using the definition of $k(h)$ we have

$$dk(h)/d(h) = (1/M(h)\,)\,(dM(h)/dh)$$

$$= k_1 + k_2 h + k_3 h^2/2! + k_4 h^3/3! + ...$$

Then, $$M(h)\, dk(h)/dh = dM(h)/dh$$

and comparing coefficient for equal powers of h, we get

$$k_1 = 1,\ k_2 = \mu_2,\ k_3 = \mu_3,\ k_4 = \mu_4 - 3\,\mu_2^2, \text{ etc.}$$

Kurtosis is defined as $k_4/(k_2)^2 = (\mu_4/(\mu_2)^2) - 3$.

Another way is to represent a distribution through its characteristic function

$$C(h) = \sum_j e^{ihxj} p(x_j) = \int_{-\infty}^{\infty} (e^{ihx}) f(x)dx$$

$C(h)$ always exists as a complex number for which $p(X_j)$ or $f(x)$ are defined, whereas $M(h)$ or $k(h)$ may or may not exist. We easily obtain, as in the case of $M(h)$, $C(h)$ as :

$$C(h) = 1 + ih = \mu_1' + h^{2}/2\,!\,(\mu_2') + \ldots..$$

or use $$C(h) \text{ as } 2h\, f(x) = \int_{-\infty}^{\infty} e^{ith}\, C(h)\, dh$$

If $x_1, x_2, \ldots x_N$ are pairwise independent random variables and L is a linear combination given by $L = \sum_j c_j x_j$, then

$$V(L) = \sum_j c_j^2 v(x_j) + \sum_{j \neq k} c_j c_k c(x_j, x_k)$$

where $$\text{cov}(x_j, x_k) = E[(x_j - \mu_j)(x_k - \mu_k)]$$

If all j, k are pairwise independent then the cross product forms vanish (uncorrelated) to yield $V(L) = \Sigma_j c_j^2 V(x_j)$. However, pairwise independence is not necessarily full or complete independence.

The Pearson correlation coefficient between random variables x_j and x_k is

$$\rho_{jk} = (x_j, x_k)/(V(x_j)\; V(x_k))^{1/2}$$

which is a non-dimensional number ranging from -1 to +1 and is zero when x_j, x_k are uncorrelated (or independent).

Therefore, $$V(L) = \Sigma_j c_j^2 \alpha_j^2 + \Sigma_{j \neq k} c_j c_k \rho_{jk} \sigma_j \sigma_k$$

It is also true that $\rho_{jk} = \rho_{kj}$.

Markov's Inequality

If X is a non-negative random variable with expectation μ and λ is any real positive number, then $P(X \geq \lambda) \leq \mu/\lambda$.

Chebyshev's inequality is deduced from Markov's inequality. If X is having first and second moments

$$E(X) = \mu \text{ and } V(X) = \sigma^2$$

Then, $$P(|X - \mu| \geq \lambda) \leq \sigma^2/\lambda^2$$

Joint distribution of two random variables

Let $h(x) = \int_{-\infty}^{\infty} f(x, y)\, dy$ and $h(y) = \int_{-\infty}^{\infty} f(x, y)\, dx$

The variables x, y are independent if $f(x,y) = f(x)\, f(y)$ or in terms of distribution function

$$F(x,y) = G(x)\, H(y)$$

where $G(x) = \int_{-\infty}^{x} 2(u) du$ and $F(x,y) = \int_{-\infty}^{x} \int_{-\infty}^{y} f(u,v) du dv$

Convolution

Let $Z = X + Y$ and $P(z)$ is distribution of z i.e. $P(x + y) \leq z$.

Hence, $P(z) = \int_{-\infty}^{\infty} g(z-y)\, h(y) dy$

or by differentiating with respect to z we get

$$p(z) = \int_{-\infty}^{\infty} g(z-y)\, h(y) dy$$

which is called the convolution of density functions g and h. It is the density function for z.

For k variates to be independent we must have

$$F(x_1 \,.....\, x_{k)} = F_1(x_1)\, F_2(x_2) \,.....\, F_k(x_k)$$

or $f(x_1 \,.....\, x_k) = f_1(x_1)\, f_2(x_2) \,.....\, f_k(x_k)$, therefore k independent samples from the same population we have $f(x_1 \,.....\, x_k) = f_1(x_1)\, f_2(x_2) \,.....\, f_k(x_k)$ since the density function f is the same for all k samples taken from the population.

2.2 Distribution of Discrete Random Variables

The main important discrete distributions are binomial (Bernoulli), Poisson and Hypergeometric, whereas among continuous univariate distributions : Gaussian(normal), rectangular, Exponential, Gamma, Beta are most important. In addition, we can get sampling distributions such as chi-square, t and F - distribution.

Binomial (Bernoulli) Distribution

$$b(x, N, \theta) = N_x\, \theta^x (1 - \theta)^{N-x}$$

We get
$$M_j(h) = E (\exp [hI_{Aj}])$$
$$= \Sigma (\exp [hI_{Aj}])\, P(I_{Aj})$$
$$= e^0 (1 - \theta) + e^h \theta$$

Therefore, $M_j(h) = (1 - \theta) + e^h \theta$ for a single trial.

So, MGF for N trials is

$$M(h) = [M_j(h)]^N = [(1 - \theta) + e^h]^N$$

The cumulant generating function is

$$\begin{aligned} k(h) &= \ln M(h) \\ &= N \log(1 - \theta) + e^h\theta \\ &= N \ln (1 - \theta h + \theta h^2/2\,! + \theta h^3/3\,! + ...) \end{aligned}$$

First four cumulants are

$$\begin{aligned} k_1 &= N\theta = \mu, \\ k_2 &= N\theta (1 - \theta) = \sigma^2 \\ k_3 &= N\theta (1 - \theta)(1 - 2\theta) = \sigma^2 (1\text{-}2\theta) \\ k_4 &= N\theta (1 - \theta)(1 - 6\theta + 6\theta^2) \\ &= \sigma^2 (1 - 6\theta + 6\theta^2) = \sigma^2 (1 - (6\,\sigma^2/N)) \end{aligned}$$

The skewness is $k_3/(k_2)^{3/2} = (1 - 2\theta)/\sigma$ and is zero only when θ is 1/2.

The *kurtosis* is $b_2 = k_4/(k_2)^2 = (1/\sigma^2) - (6/N)$.

If θ is 1/2, $\sigma^2 = N/4$ and $k_3/(k_2)^{3/2} = 0$ and $k_4/(k_2)^2 = -2/N$

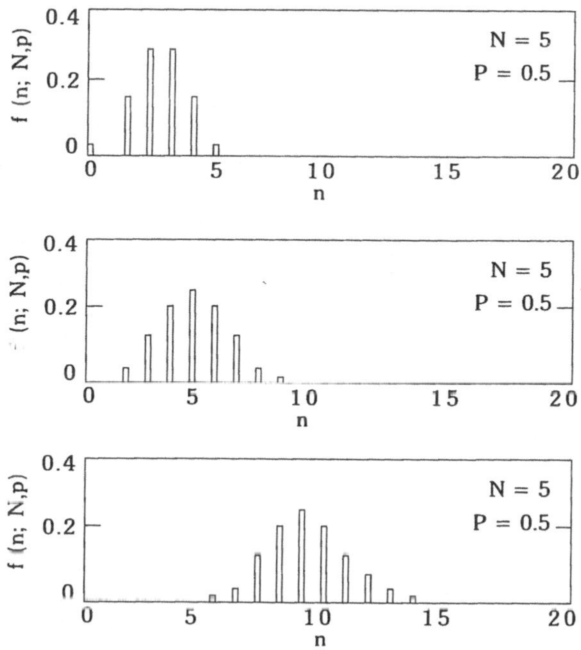

Fig. 2.1 Binomial distribution with various N and p = 1/2

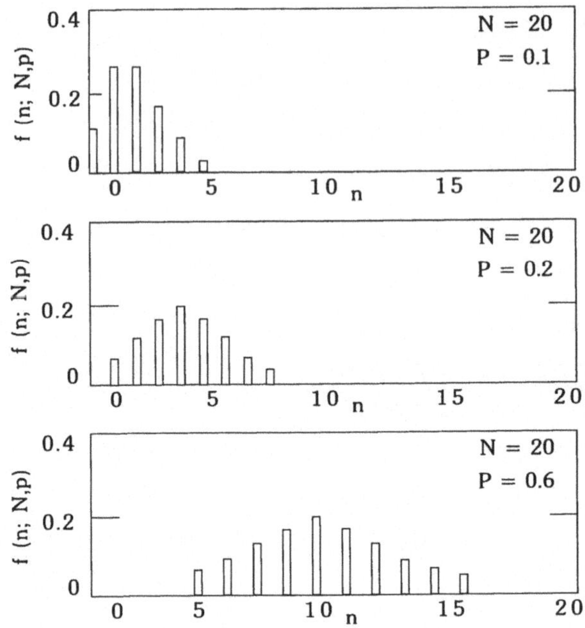

Fig. 2.2 Binomial distribution with various p values for fixed N = 20

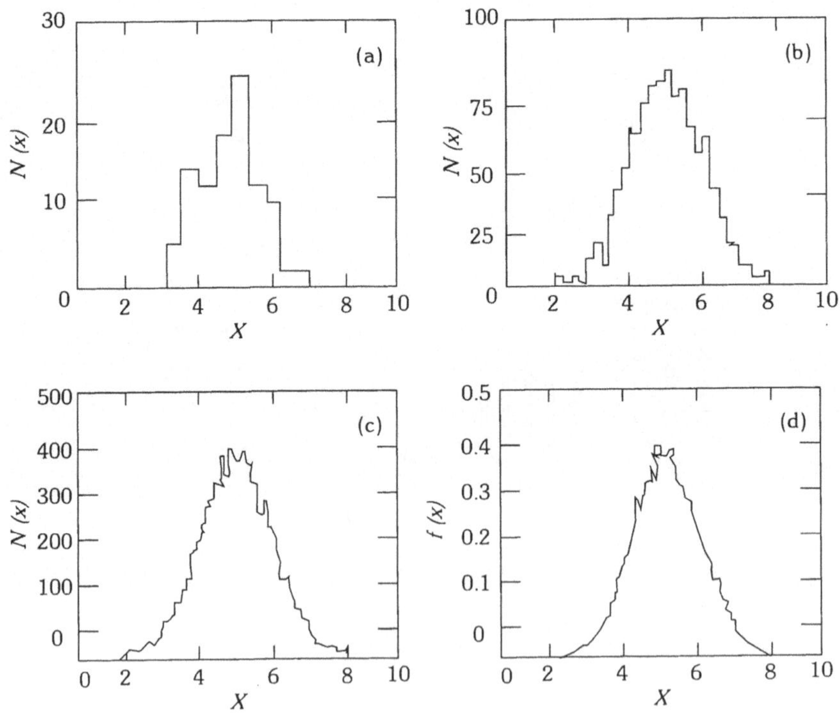

Fig. 2.3 Histograms with various p.d.f.s with different Δx

r^{th} moment $\mu_r = \Sigma_{x=0}^{N} \binom{N}{x} \theta^x (1-\theta)^{N-x} (x-N)^r$ yield recursion formula for

the moments

$$M_{r+1} = \theta(1-\theta)\,[N_r\,\mu_{r-1} + d\mu\,r/d\,\theta]$$
$$K_{r+1} = \theta(1-\theta)\,dK_r/d\,\theta,\ r \geq 2;$$

so $k_1 = N\theta$ and higher moments are easily obtained. (Fig. 2.1, 2.2 and 2.3).

Weak Law of Large Numbers

Let P_N be the relative frequency of success in N independent trials with success probability given by θ being same for all trials. Since $E(P_N) = \theta$, we have by Chebyshev inequality

$$P(|P_N - \theta| \leq \lambda) \leq \sigma^2/\lambda^2$$

Here $\sigma^2 = \theta(1-\theta)/N$ and θ lies between 0 and 1, $\theta(1-\theta) = 1/4$.

So the above inequality reduces to

$$P(|P_N - \theta| \leq \lambda) \leq 1/4N\,\lambda^2$$

For any fixed $\lambda = 0$, and any given $E > 0$, we can always find N such that $1/(4N\,\lambda^2) < E$. This means that for a large enough N the difference between P_N and θ can be made as close to zero as we have. This is to say P_N converges in probability to θ. This convergence in probability is not same as mathematical convergence.

In non-Bernoulli sampling the probability of success θ_j may vary with the trial j (Poisson scheme) or the probability of success may depend on previous trials (hypergeometric scheme).

For Poisson's scheme, we have

$$\mu = E(X) = \Sigma_j P(A_j) = \Sigma\theta_j$$

also $$\sigma^2 = \Sigma_j\ V(IA_j) = \Sigma_j (\theta_j - \theta_j)^2.$$

Defining $\theta = 1/N \sum \theta_1$ and $= 1/N(\theta_j - \theta)^2 = 1/N\ \Sigma_j\ \theta_j^2 - \theta^2$ we obtain

$$\sigma^2 = N\theta - N(\sigma_\theta^2 + \theta^2) = N\theta(1-\theta) - N\sigma_\theta^2$$

So, in Poisson's scheme the variance is less than that of binomial scheme.

In hypergeometric scheme $h(x, N, M, S) - (S_x)(F_{N-x})/(M_N)$, where $M = S + F$ and N balls are drawn from these. When M is very large and θ not near 0 or 1, this probability approximates to $b(x, N, \theta)$, with same algebra we have

$$E(X) = \mu = N\theta - NS/M \text{ and}$$
$$V(X) = \sigma^2 = N\theta\,[(S-1)(N-1)/(M-1) + (M - NS)/M]$$
$$= N\theta\,(1-\theta) - (M-N)/(M-1)$$

When M is very large compared to N, the factor $(M{-}N)/(M{-}1)$ tends to 1.

Poisson distribution

If the success probability is very low (rare) but N is so large with the expected number of success in the sample is moderate (0.1 and 1.0),then probability of exact success is approximated by $P(x, \mu) = \mu^x e^{-\mu}/x\,!$, where $\mu = N\theta$. This gives Poisson's distribution for $x = 0$ to ∞. We can derive Poisson's distribution from binomial by making N tends to infinity and θ tending to zero in such a way as $N\theta$ tends to μ.

So $$\underset{N\to\infty,\theta\to 0}{\text{Limit}}\; b(x, N, \theta) = P(x, \mu)$$

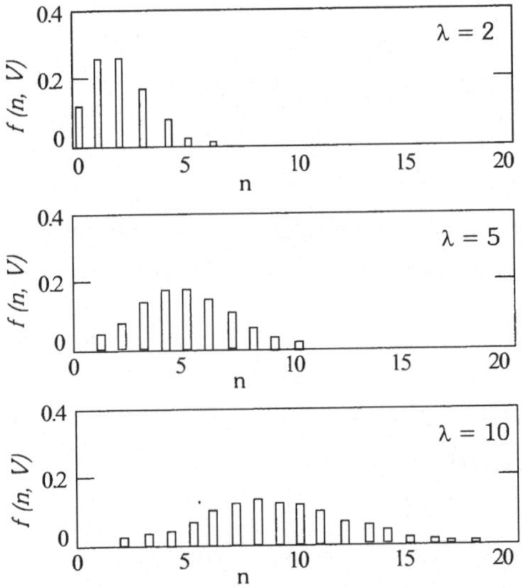

Fig. 2.4 Poisson p.d.f.s for various λ

The mean is $\mu = E(X)$ and variance $\sigma^2 = \mu$ for Poisson's distribution (Fig. 2.4).

Normal distribution as an approximation to the Binomial

Binomial distribution has mean $\mu = N\theta$ and variance $= N\theta\,(1-\theta)$ which are correlated. So instead of random variable X we can standardize it by $Z = (X - \mu)/\sigma$. With some algebra we obtain

$$\log P \approx -1/2 \log (2\pi N\theta(1-\theta) - Z^2/2 + Z/2N^{1/2}[- - -] + \ldots$$

For large N, the $N^{-1/2}$ terms may be neglected and also $N^{-1/2}$ term vanishes whenever

$$\theta = (1 - \theta) = 1/2.$$

Thus, we have

$$\log P \approx -(2\pi N\theta (1 - \theta)) - Z^2/2$$

or $$P \approx (1/\sigma \sqrt{2}\,\pi)\, e^{-(x-\mu)2/(2\sigma^2)},$$

where $$N\theta (1 - \theta) = \sigma^2$$

The limit form of the σ^P is, therefore, lim $\sigma P = (1/\sqrt{2}\,\pi) e^{-(z)2}/2$ or the standard normal gaussian distribution.

The normal distribution is an even function, $\phi(Z) = \phi(-Z)$ with maximum value of 0.3989 at $Z = 0$ and percentiles $\phi_1 = 0.25$ and $\phi_3 = 0.75$ at $Z = \pm 0.6745$

$$\text{M.g.f, } M(h) = \int_{-\infty}^{\infty} e^{hz}\, \phi(z)\, dz$$

$$= (2\pi)^{-1/2} \int_{-\infty}^{\infty} e^{hz} e^{-\frac{1}{2}z^2}\, dz$$

Substituting $h = (z - h)$ we get

$$M(h) = 1 + \frac{h^2}{2} + \frac{1}{2!}\frac{h^4}{4} + \frac{1}{3!}\frac{h^6}{6} + \ldots..$$

Mean =0, and $\mu_2 = 1$, $\mu_4 = 3$, $\mu_6 = 15$ etc.

$$\text{c.g.f } K(h) = \ln M(h) = h^2/2.$$

The only nonzero cumulant is $k_2 = 1$. (Fig. 2.5(a, b, c)).

Comparing the parameters of standard normal distribution we obtain $k_1 = \mu$ and $k_2 = \sigma^2$.

The great simplification given by cumulants is the reason for their use in statistical theory. The support for the entire distribution is within $Z = \pm 4$. We can use the cumulative frequency versus the random variable graph on the probability paper to obtain an indication whether the random variable is linear i.e. normal (gaussian), otherwise it is nongaussian. For Gaussian distribution mean and variance are independent parameters. (Fig. 2.5(a, b, c)).

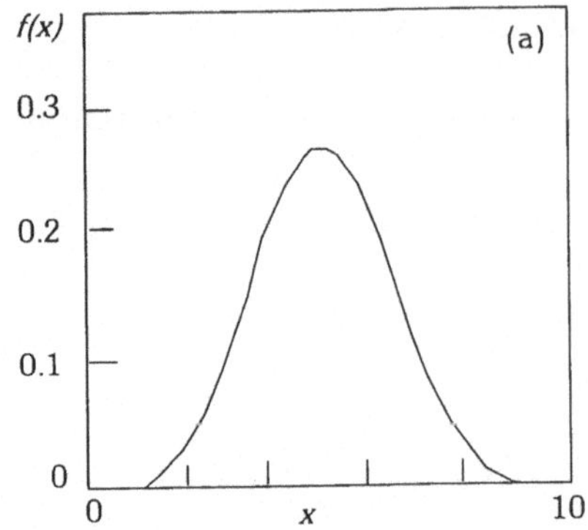

(a) Gaussian p.d.f.

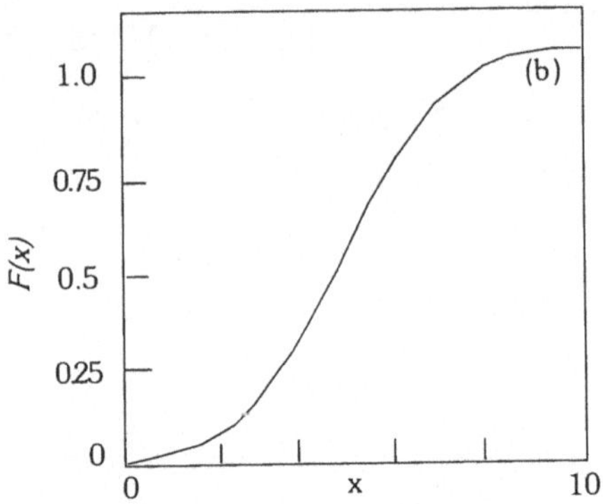

(b) Gaussian distribution c.d.f.

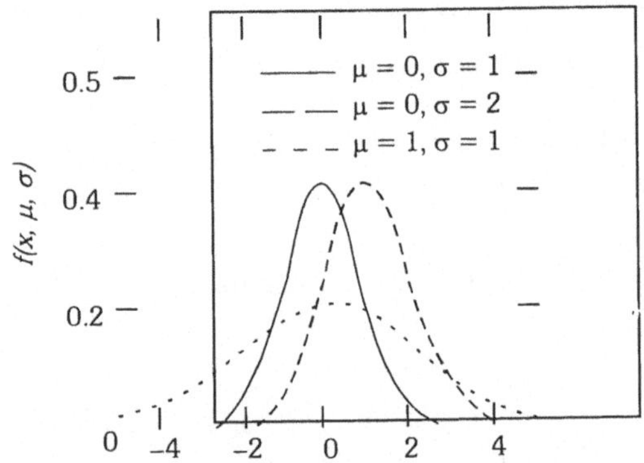

(c) Gaussian p.d.f. for various values of parameters μ and σ

Fig. 2.5

The binomial success probability $P = X/N$ can also be transformed as $A = \sin^{-1}(P^{1/2})$ to yield a normal/gaussian distribution. The key is to make mean and variance independent for the transformed variable. A small angle or degree

δA is given by

$$\delta A \approx (dA/dp)\ \delta p = (180/\pi)\ [1/2p^{1/2}\ (1 - p)^{1/2}]\ \delta p$$

$$(\delta A)^2 \approx (90/\pi)^2\ ((\delta p)^2/p(1 - p))$$

$$= (821(\delta p)^2)/(p(1 - p)); \text{ therefore,}$$

$$V(A) \approx (821/p(1 - p))\ V(p)$$

$$= (821/p(1 - p))\ (p(1 - p)/N) = 821/N$$

Thus $V(A)$ is a constant and independent of A as desired. For Poisson's distribution P is small and so $A = \sin^{-1}(P^{1/2}) = P^{1/2} = (X^{1/2}/N^{1/2})$

Therefore $Y = N^{1/2} A = X^{1/2}$ gives gaussian/normal distribution for a Poisson random variate.

2.3 Distribution of Other Continuous Random Variables

Uniform (Rectangular)

The property of success is same over the range of rectangular variable $[\alpha, \beta]$. So, $f(x) = (\beta - \alpha)^{-1}$ with mean $\mu = (\alpha + \beta)/2$ and variance $= (\beta - \alpha)^2/12$ (Fig. 2.6).

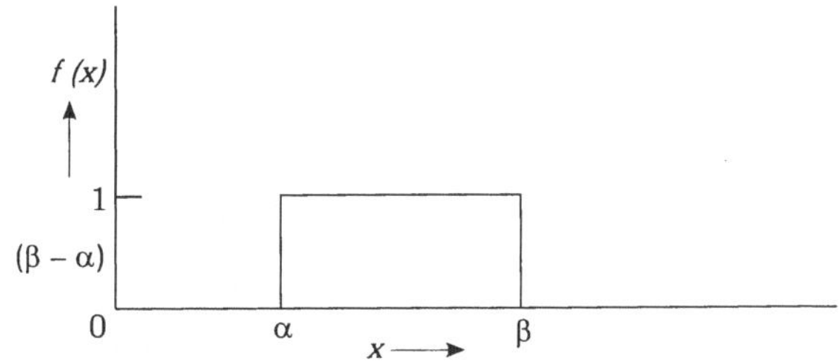

Fig. 2.6 Rectangular (uniform) p.d.f.

The variable $Y = F(x)$ has uniform distribution with range 0 to 1. This transformation is called *probability transformation.*

Gamma distribution

$$f(x) = e^{-x} x^{\alpha-1}/\gamma(\alpha),\ 0 \le x < \infty \text{ with } \alpha > 0 \text{ and } \gamma(\alpha + 1) = \alpha\, \sigma(\alpha). \text{ (Fig. 2.7(a, b, c)).}$$

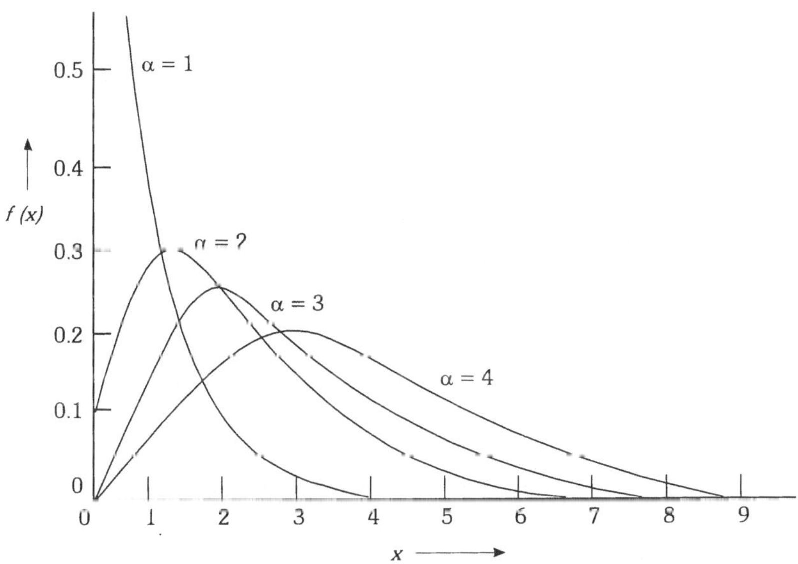

(a) Gamma p.d.f. for different values of parameter α

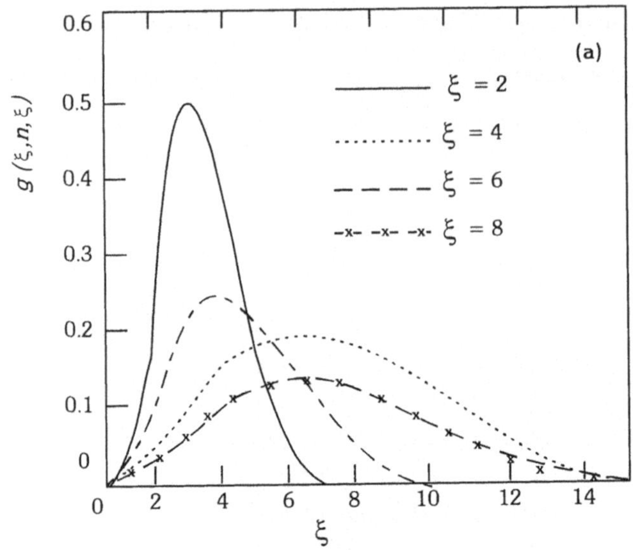

(b)

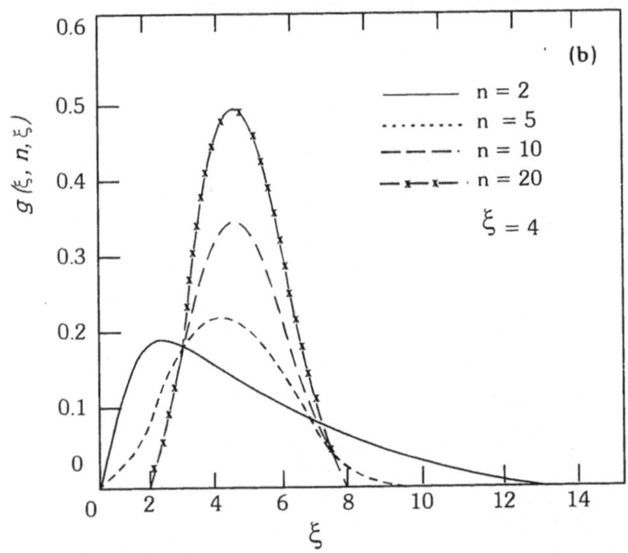

(c)

Fig. 2.7 The sampling p.d.f. for estimators $\hat{\xi}$ for various values of n and ξ for gamma distribution

If $\alpha = 1$ it gives an exponential distribution (Fig. 2.8(a, b)) and $\alpha = 1/2$ gives the standard gaussian distribution.

The m.g.f., $\quad M(h) = \int_0^{\infty} e^{hx} f(x)dx = (1 - h)^{-\alpha}.$

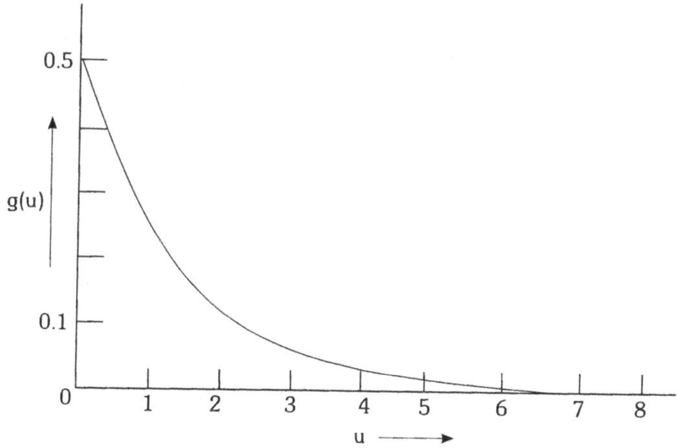

(a) Exponential p.d.f.

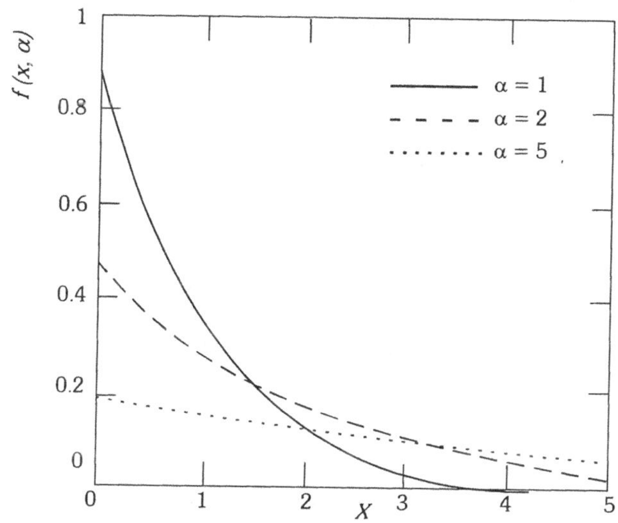

(b) Exponential distribution with different α values

Fig. 2.8

The c.g.f. $K(h) = \ln M(h) = -a \ln(1 - h)$

$$= \alpha\,(h + h^2/2 + h^3/3 + \ldots.),$$

so that the first few cumulants are

$$k_1 = \alpha,\ \ k_2 = \alpha,\ \ k_3 = 2\alpha$$
$$k_4 = 6\alpha \text{ and } k_r = \rho\,\gamma\,(r)$$

The skewness is $k_3/(k_2)^{3/2} = 2\alpha^{1/2}$, which is always positive. The kurtosis is $b_2 = k_4/(k_2)^2 = 6\alpha^{-1}$. It has only one parameter mean = variance = α and it becomes nearly symmetrical as α increases.

We know that c.g.f of sum of independent variables can be obtained by sum of individual c.g.fs. Therefore, the sum of n independent gamma variables with parameters $\alpha_1, \alpha_2, \ldots.. \alpha_n$ has the c.g.f. $K(h) = -\sum \alpha_i \ln(1-h)$ and this is the c.g.f. of a gamma variate with parameter $\sum_j \alpha_j$. The sum of independent gamma gives a new gamma distribution which has important applications in sampling distribution and tests of hypothesis (χ^2 and F Tests).

Beta distributions

The two parameter distribution

$$F(x) = f(x) = x^{a-1}(1-x)^{\beta-1}/B(\alpha, \beta),\ 0 \le x \le 1,$$

where $a > 0,\ b > 0$

and $B(\alpha, \beta)$ beta function $= \Gamma(\alpha)\ \Gamma(\beta)/\Gamma(\alpha + \beta)$

The rth moment is $\mu_r' = \Gamma(\alpha + \gamma)\ \Gamma(\alpha + \beta)/(\alpha)\ (\alpha + \beta + \gamma)$ (Fig. 2.9).

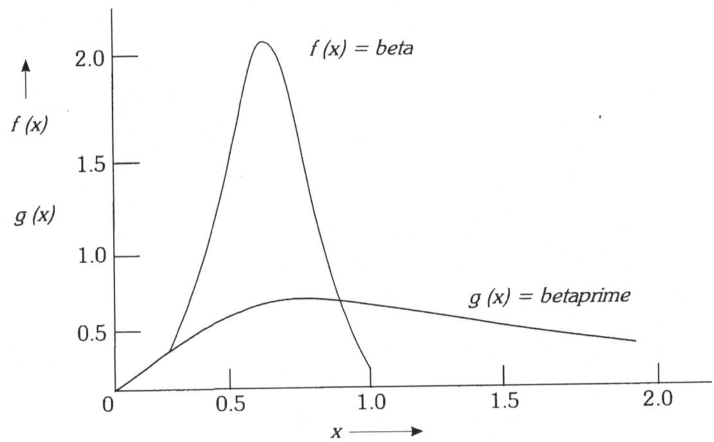

Fig. 2.9 Beta and Betaprime p.d.f.

Chi-square distribution

Let $x_1 \ldots.. x_n$ be n independent normal variables with means $\mu_1, \mu_2 \ldots.. \mu_n$ and variances $\sigma_1, \sigma_2 \ldots.. \sigma_n$. The standard variate corresponding to x_i is

$$Z_i = (X_i - \mu_i)/\sigma_i.$$

We know that $1/2\ z_j^2$ has the gamma distribution with parameter $\alpha = 1/2$, so that $\sum_{j=1}^{n} 1/2z_j^2$ is a gamma variate with parameter $\sum\alpha$, in this case $n/2$. (Fig. 2.10(a,b)). Putting $\sum z_j^2$ by χ^2 the variate $\chi^2/2$ has gamma distribution as

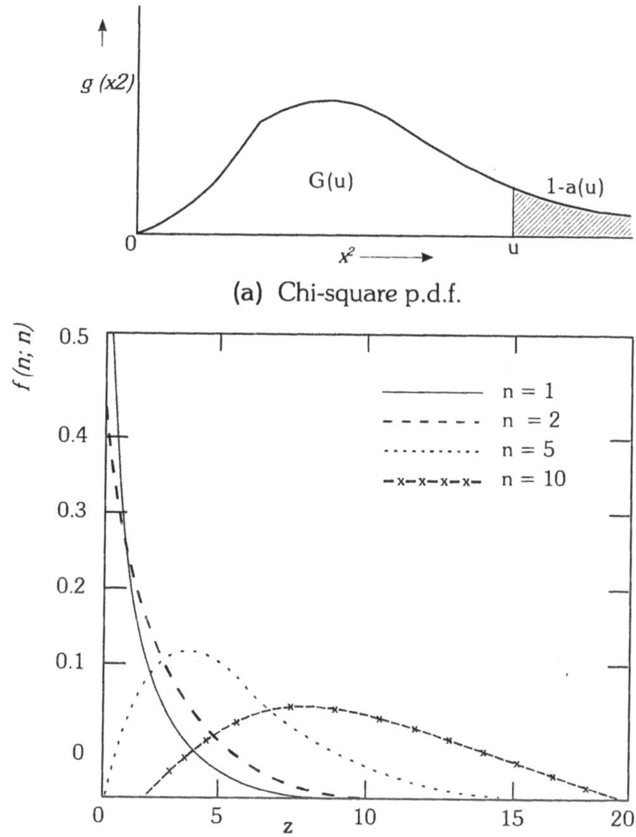

(b) Chi-square p.d.f. for various d.f. n

Fig. 2.10

$$\Gamma(\chi^2/2) = (\chi^2/2)^{((n/2)-1)}\, e^{-(\chi^2/2)}/\Gamma(n/2)$$

The distribution of χ^2 is given by $g\ (\chi^2)$ as

$$g\ (\chi^2)\ d\ (\chi^2) = f\ (\chi^2/2)\ d\ (\chi^2/2) \text{ or}$$

$$g(\chi^2) = (\chi^2/2)^{((n/2)-1)}\, e^{-(\chi^2/2)}/\Gamma(n/2)$$

The rth cumulants are given by $2^r\ k_r = 2^{r-1}(r-1)!n$

Thus $\quad k_1 = E\ (\chi^2) = n,\ k_2 = V(\chi^2) = 2n;\ k_3/(k_2)^{3/2} = 2(2/n)^{1/2}$,

distribution function $G(u) = \int_0^a g\ (\chi^2)\ d\ (\chi^2)$

Log - normal distribution

Many geological variables are lognormal. For example grain size distribution, trace components, permeability distribution, bed thickness distribution etc. (Fig. 2.11).

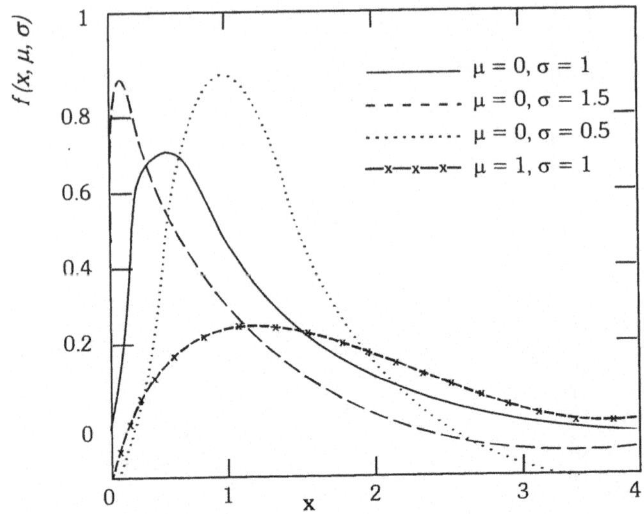

Fig. 2.11 Log-normal p.d.f. for several values of parameters μ and σ

Let $Y = \log X$ be normally distribution with mean α and variance β. It is obvious X has a range $(0, \infty)$ so the lognormal distribution is applicable.

Then, $$f(x)\, dx = g(y)\, dy = g(y)\,(1/x)dx \text{ or } (y-\alpha)^2/2\beta$$

$$f(x) = x^{-1}\, g(y) = x^{-1}\,(2\pi\beta)^{-1/2}$$

The rth moment about origin of variate X is

$$\mu_r' = \int_0^\infty x^r f(x)\, dx = \int_0^\infty e^{ry} g(y) dy = \exp\,(r\alpha + r^2\,\beta/2)$$

The mean of X is ,therefore,

$$\mu_r' = \mu = \exp(\alpha + \beta/2)$$

and $$\sigma^2 = \mu_2^1 - (\mu_2^1)^2 \exp\,(2\alpha + 2\beta) - \exp\,(2\alpha + \beta) = \mu^2\,\eta^2,$$

where $$\eta^2 = e^\beta - 1.$$

The coefficient $\eta = \sigma/\mu$ is called *coefficient of variation.*

The skewness is $r_1 = \eta^3 + 3\eta$.

Central Limit Theorem

Let x_1 ..... x_n ..... be independent random variables all having same distribution with mean μ and finite variance $\sigma^2 < \infty$, but not necessarily gaussian. The standardized variate for x_j is

$$Z_j = (x_j - \mu)/\sigma$$

and we define $Yn = \sum z_j/n^{1/2} = n^{1/2} z,$

where Z is the arithmatic mean of the Z_j.

Then, $n \to \infty$, Yn tends to standard normal variate i.e. $P(Yn < y) \to \phi(y)$.

Applying the central limit theorem to n samples for a given normal population with mean μ and finite variance σ^2 we get,

$$f(\bar{x}) = f\left(\textstyle\sum_{j=1}^{n} x_j/n\right) = \varphi[(\bar{x} - \mu)/(\sigma/n)]$$

This is an important result for tests of hypothesis.

2.4 Sample and Population

In most statistical problems samples are drawn from a parent population having parameters defining the population distribution. Sample statistics gives concise description of samples but these are more importantly used for decisions regarding population parameters. The normal distribution has two parameters, population mean and population standard deviation. These parameters may be estimated from random sample characteristics such as median, sample range etc.

As the sample size (N) increases the sample statistics is expected to provide greater accuracy and precision for estimation of population parameters. We can study the distribution of sample statistics about the corresponding population parameters and obtain not only point estimates but also confidence limits to the concerned population parameter. For example, confidence interval for population mean for sample having size N is given by $\bar{x} + 1.96\,\sigma/\sqrt{N}$, if σ is known. If σ is unknown it has to be estimated as sample standard deviation.

Similarly we can obtain precision and internal estimates for binomial distribution. Population success probability $\theta = \sum x_j / N$ and $V(X) = N\theta(1 - \theta)$. The lower and upper 95% confidence limits are $X \pm 1.96\, X^{1/2} (1 - (X/N))^{1/2}$ for binomial data provided N is large.

2.5 Change of Variables (Fig. 2.12(a, b))

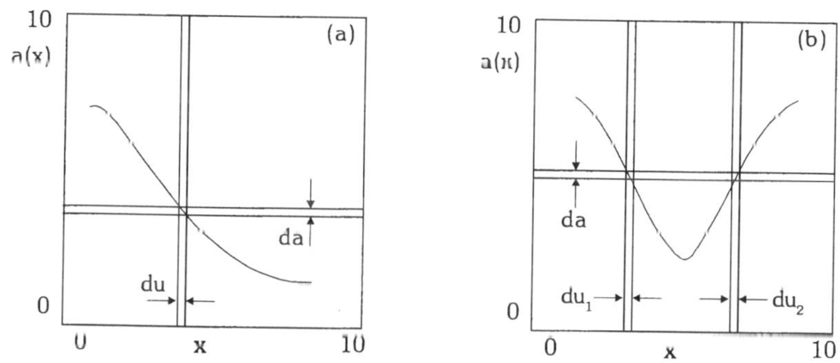

(a) 1 : 1 transformation of variable

(b) Transformation of variable with multiple regions R_i and R_i'

Fig. 2.12

What is the probability distribution of y when probability distribution of random variable x is known ? We know, $y = f(x)$ when both x, y are continuous with density functions $h(x)$ and $g(y)$. Assuming f is monotonic we obtain

$$g(y) = f[f^{-1}(y)] \,.\, |df^{-1}/dy| = f(x) / \,|\, dy/dx \,|$$

where $|df^{-1}/dy|$ is called the *Jacobian of transformation.*

Often we want $g(y)$ to be normally distributed, where f is normalizing transformation. If f is multivalued then

$$g(y) = \Sigma_{j=1}^{m} \; h\,[x_i = \psi_i\,(y)]/|\; f\,(x)|$$

Similarly if $f(x_1,x_2)$ be transformed to $g(y_1, y_2)$ where $y_1 = a_{11}\,x_1 + a_{12}\,x_2$, $y_2 = a_{21}\,x_1 + a_{22}\,x_2$ then, $g\,(y_1, y_2) = f\,(b_{11}\,y_1 + b_{12}\,y_2, b_{21}\,y_1 + b_{22}\,y_2) \,/\, |a_{11}\,a_{22} - a_{12}\,a_{22}|$.

The absolute value of the determinant appearing as denominator is called the Jacobian of the transformation where x_1, x_2 are obtained as $x_1 = b_{11}\,y_1 + b_{12}\,y_2$, $x_2 = b_{21}\,y_1 + b_{22}\,y_2$. The distribution of normalized variable y given by density function g or cumulative distribution G can be tested for normality by linearity of cumulative frequency plot on probability graph paper, chi-square test for the observed frequency versus expected frequency or by the skewness and kurtosis values for g compared to normal values of 0 and 3, respectively.

Suppose the transformation is not one-to-one but the range of x can be divided into finite number of exclusive regions R_1, ... R_k with the corresponding regions R_1' ..... R_k' (may be overlapping) in the range of Y such that the transformation from R_i to R_i' is one to one.

Let $x = g_i\,(y)$ be the transformation from to R_i' to R_1 and let $g_i'\,(y)$ be continuous.

Then the density at y is

$$\Sigma_{j=1}^{k} \; \pi_i\, p_x\; [g_i\,(y)]\; |g_i'\,(y)|$$

where $\pi_i = 1$ if $y \in R_i'$ and $\pi_i = 0$, otherwise.

For instance if $Y = X^2$, there are two regions to be considered : $R_1 = (-\infty, 0)$ and $R_2 = (0, \infty)$. The corresponding regions in the space Y are $R_1' = (0, \infty)$ and $= (0, \infty)$ which are completely overlapping. The inverse functions are

$x = -\sqrt{y}$ corresponding to (R_1, R_1') $x = \sqrt{y}$ corresponding to (R_2, R_2').

Every $y \in R_1'$ and R_2', therefore the p.d.f at y is

$$\left[p_x(\sqrt{y}) + p_x(-\sqrt{y})\right]\left[\frac{1}{(2y)^{-1/2}}\right]$$

These results can be extended to several variables in an analogous manner.

2.6 Examples

Example 1

Find the expectation and variance of the number of spots obtained in rolling a dice.

Answer :

$$E(X) = \sum_{j=1}^{6} jx\frac{1}{6} = \frac{1}{6}x21 = 3.5$$

$$V(X) = \sum_{j=1}^{6} j^2x\frac{1}{6} - (E(X))^2 = \frac{1}{6}x91 - [\frac{7}{2}]^2 = 2.917$$

The standard deviation $\sigma = \sqrt{2.917} = 1.71$

Example 2

Using moment generating function, find the mean and variance of rectangular distribution ranging from -a to +a.

Answer :

$$f(x) = \frac{1}{2a} \text{ for } x \in [-a, +a] \text{ and zero elsewhere.}$$

$$M(h) = \frac{1}{2a}\int_{-a}^{a} e^{hx}dx = \frac{1}{2ah}(e^{ah} - e^{-ah}) = \frac{1}{ah}\sinh ah$$

Expanding in a power series

$$M(h) = 1 + \frac{a^2h^2}{3!} + \frac{a^4h^4}{5!} + \ldots\ldots$$

So all odd moments are zero and even moments are given by

$$\mu'_2 = \frac{a^2}{3},\ \mu'_4 = \frac{a^4}{5} \text{ etc.}$$

Hence, mean = zero and $\mu'_r = \mu_r$

$$\text{Variance} = \mu_2 - \mu'_2 = \frac{a^2}{3} = \frac{4a^2}{12} = \frac{(range)^2}{12} = \frac{R^2}{1^2}$$

Example 3

Find the variance of sample skewness and kurtosis for Normal populations.

Answer :

For the third cumulant

$$E(k_3) = 0 \text{ and } V(k_3) = 6k_2^3\frac{N}{(N-1)(N-2)}$$

And an unbiased estimator

$$V(k_3) = 6k_2^3 \frac{N(N-1)}{(N-2)(N+1)(N+3)}$$

For the fourth cumulant

$$E(k_4) = 0, \; V(k_4) = 24k_2^4 \frac{N(N-1)^2}{(N-3)(N-2)(N+3)(N+5)}$$

The variance for $g_1 = \frac{k_3}{k_2^{3/2}}$ and $g_2 = \frac{k_4}{k_2^2}$ are

$$V(g_1) = \frac{6N(N-1)}{(N-2)(N+1)(N+3)} \text{ and}$$

$$V(g_2) = \frac{24N(N-1)^2}{(N-3)(N-2)(N+3)(N+5)}$$

For large N the values are approximately $\frac{6}{N}$ and $\frac{24}{N}$, respectively.

Example 4

Prove that sample correlation coefficient r varies from –1 to +1. (Fig. 2.13)

Answer :

Let the normal random variables be designated X (predictor), Y (criterion).

Hence, standardized variables Z_y, Z_x have expectations zero, Variances 1, Correlations (r).

The linear computate $L = Z_y \pm Z_x$ has expectation $E(L) = 0$ and

$$V(L) = E(L^2) = Z_y^2\, 2\, Z_y\, Z_x + \geq 0$$

or $$2(1.0 \pm r_{xy}) \geq 0.$$

This simplifies to $1 \pm r_{xy} \geq 0$, giving $-1 \leq r_{xy} \leq +1$.

We can also obtain Covariance $(X, Y) = r_{xy}\, \sigma_y\, \sigma_x$, for the definition of Correlation Coefficient Fisher's z' transform, $z' = \frac{1}{2}\log\left[\frac{1+r}{1-r}\right] = \tanh^{-1} r$ having

$E(z') = \frac{1}{2}\log_e\left[\frac{1+\rho}{1-\rho}\right](=\zeta), V(z') = \frac{1}{(N-3)}$, simultaneously achieves both normalization as well as stabilization of its variance.

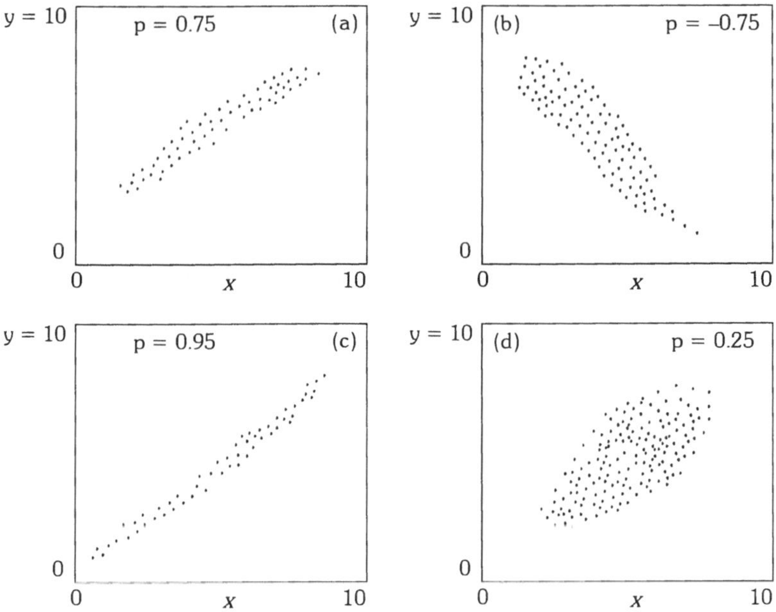

Fig. 2.13 Correlation between *y* and *x* having various magnitudes

Example 5

Normalize a binomial variate $p = \dfrac{X}{N}$

Answer :

The criterion : mean and variance of transformed variable are independent.

Let the normalizing transform be $A = \sin^{-1}(p^{1/2})$.

Then, $$\delta p \approx \frac{dA}{dp}\,\delta p = \frac{180}{\pi}\,\frac{1}{2p^{1/2}(1-p)^{1/2}}\,\delta p.$$

Hence, $$(\delta A)^2 \approx \left[\frac{90}{\pi}\right]^2 \frac{(\delta p)^2}{p(1-p)} \approx \frac{821(\delta p)^2}{p(1-p)} \approx \frac{821V(p)}{p(1-p)}$$

We know, $$V(p) = \frac{p(1-p)}{N}$$

So, $$V(A) = (\delta A)^2 \approx \frac{821}{N}$$ which is independent of variable p.

For Poisson distribution p tends to zero, so normalizing transform $\sin^{-1}\sqrt{p}$ reduces to $\sqrt{p}$.

Example 6

What is the probability of the largest observations in sample size N from a unit Normal population being less than x ?

Answer :

It is given by $p(x_n \leq x) = [\Phi(x)]^N = (2x)^{-1/2} \int_{-\infty}^{x} e^{-\frac{z^2}{2}} dz$.

For the smallest observation, tabular values below should be negative and Upper changed to Lower.

Sample Size N	Upper percentage, 5%	Points 1%
5	2.319	2.877
10	2.568	3.089
15	2.705	3.207
20	2.799	3.289
30	2.929	3.402
50	3.082	3.539
100	3.283	3.718

Outliers may be rejected by computing $z_i = (x_i - \bar{x})/s$ by t-test with $(N-2)$ degrees of freedom for a single observation. For $N = 20$, t - value for one rejection corresponding to 0.05 and so on.

The statistics needed for $N > 1$ is $t_i = \left[\frac{N-2}{\left((N-1)^2/N\right) - z_i^2}\right]^{1/2}$

Example 7

What is the density function of range (R) ?

Answer :

Range $= x_N - x_1 = R$

$$g(R) = N(N-1) \int_{-\infty}^{\infty} [F(x_1 + R) - F(x_1)]^{N-2} f(x_1) f(x_1 + R) dx_1,$$

$$E(R) = \int_{-\infty}^{\infty} [1 - F^N - (1-F)^N] dx,$$

which is close to 6 for sample sizes $N = 350$ to 650.

Therefore, standard deviation, S = range/6 = $R/6$.

For small sample sizes $N \leq 10$, we approximately have $\sigma_R = \frac{R}{E(R)} = kR$ or $k = \frac{R}{\sigma_R}$. The lower and upper 0.5% values to compute 99% confidence limits in a single sample is also given below along with k values.

N	k	Lower 0.5%	Upper 0.5%
2	.886	.01	3.97
3	.591	.13	4.42
4	.486	.34	4.69
5	.430	.53	4.89
6	.395	.75	5.03
7	.370	.92	5.15
8	.351	1.08	5.26
9	.337	1.21	5.39
10	.325	1.33	5.42
15	.288	1.80	5.70
20	.268	2.12	5.89

Example 8 : (Order statistics)

x_r is r^{th} order statistics.

$$g(x)dx = c[F(x)]^{r-1}[1 - f(x)]^{N-r} f(x)dx$$

where $\quad c^{-1} = B(r, N - r + 1)$

By using Normal distribution for the population $c^{-1} = B(r + 1, r + 1) = (r!)^2/N!$ Expected value of median is zero (by symmetry of Normal population) and

$$V(x_{r+1}) = \frac{\pi}{2(N+2)} + \frac{\pi^2}{4(N+2)(N+4)}$$

As N increases, the efficiency of median x_{r+1} become $2/\pi = 0.637$ (i.e. 64% compared to arithmetic mean $\bar{x}$. Hence, arithmetic mean is better estimator of population mean as compared to median,

Example 9 : Change of variables (Fig. 2.14(a, b)

Suppose $f(x) = 1$ for $0 < x < 1$ and zero otherwise.

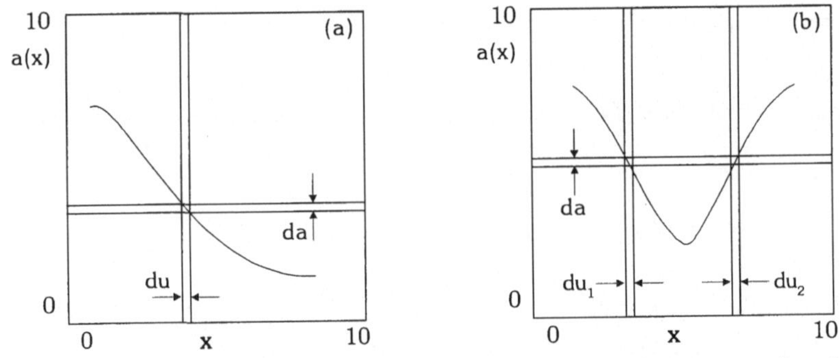

(a) 1 : 1 transformation of variable

(b) Transformation of variable with multiple regions R_i and R_i'

Fig. 2.14

Let $\quad Y = X^2$ obtain density $g(y)$ of Y

We have $\quad \dfrac{dy}{dx} = 2x,$

So, $\quad g(y) = \dfrac{f(x)}{\left|\dfrac{dy}{dx}\right|} = \dfrac{1}{2x}, 0 < x < 1,$

or $\quad g(y) = \dfrac{1}{2\sqrt{y}}, 0 < y < 1$

We check the integral $\int_0^1 \dfrac{1}{2\sqrt{y}} dy = [y]_0^1 = 1$

Alternatively, we have $G(y) = P(Y<y)$

$$= P(X < \sqrt{y})$$

$$= \int_0^{\sqrt{y}} f(x)dx = \int_0^{\sqrt{y}} dx = \sqrt{y}$$

So, we have $\quad g(y) = \dfrac{gG(y)}{dy} = \dfrac{1}{2\sqrt{y}}$, as before.

The second approach as well as graphical methods are more fundamental and always give correct results. Note that supports of a density g is the range (y_1, y_2) over which it is positive.

Example 10 : *Mixture random variable (Discrete cum continuous)*

Let X be a mixture random variable taking discrete values x_i, $i = 1, 2, 3, \ldots n$ with probability p_i and a continuum of values in the interval $[a, b]$ according to density $f(x)$,

i.e. $$\text{if } [a, b] \supset [x_1, x_2],\ P(x_1 \leq X \leq x_2) = \int_{x_1}^{x_2} f(x)dx$$

Then $$EX = \sum_{i=1}^{n} x_i p_i + \int_a^b xf(x)dx$$

with the constraint

$$\sum_{i=1}^{n} p_i + \int_a^b xf(x)dx = 1$$

Thin section size analysis form a mixture distribution

Example 11 : *pdf and characteristic functions for some common probability distributions*

Distribution	pdf	Characteristic function $\phi(K)$
Binomial	$f(x,N,p) = \binom{N}{x} p^x (1-p)^{N-x}$	$\left[p(e^{ik}-1)+1\right]^N$
Poisson	$f(x,v) = \frac{v^x}{x!} e^{-v}$	$\exp\left[v(e^{ik}-1)\right]$
Uniform	$f(x,\alpha,\beta) = \frac{\Lambda}{(\beta-\alpha)}, \alpha \leq x \leq \beta$	$\frac{e^{i\beta k} - e^{i\varepsilon k}}{(\beta-\alpha)ik}$
Exponential	$f(x,\alpha) = \frac{1}{\alpha} e^{-x/\alpha}$	$1/(1 \quad ik\alpha)$
Gaussian	$f(x,\mu,\sigma^2) = \frac{1}{\sqrt{2\pi\sigma^2}} \exp\left(\frac{-(x-\mu)^2}{2\sigma^2}\right)$	$\exp\left(i\mu k - \frac{1}{2}\sigma^2 k^2\right)$
Chi-square	$f(z,n) = \frac{1}{2^{n/2}\sqrt{n/2}} z^{n/2} e^{-z/2}$	$(1-2ik)^{-n/2}$
Cauchy	$f(x) = \frac{1}{Z}\frac{1}{1+x^2}$	$e^{\lvert k\rvert}$ (Fig. 2.15)

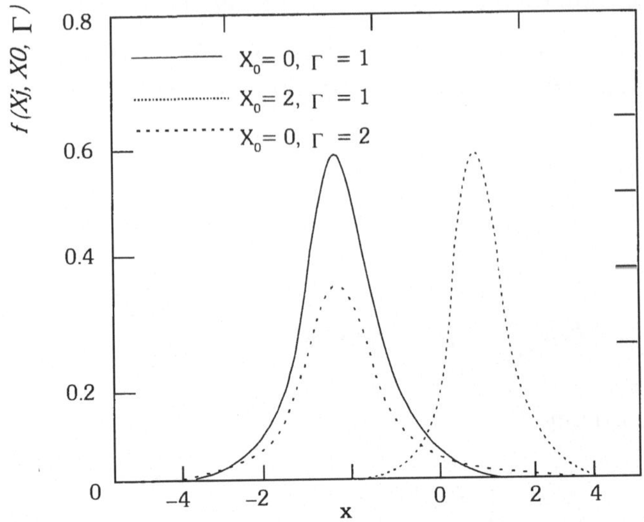

Fig. 2.15 Cauchy p.d.f. for various values of parameters x_0 and r

Example 12 : Proof of Central Limit Theorem using characteristic function

Statement:

For large n the sum $\Sigma_j\ x_j$ becomes a gaussian random variable with mean $\Sigma_j\ \mu_j$ and variance $\Sigma_j\ \sigma_j^2$ regardless of distributions of x_j.

Regarding n to be fixed for the moment we obtain

$$y_j = (x_j - \mu_j)/\sqrt{n} \text{ so that}$$

$$E\left|y_j\right| = 0 \text{ and } E\left|y^2{}_j\right| = \sigma_j^2/n$$

If all variables σ_j^2 are equal then the characteristic function ϕ, (K) for y_i can be expanded in a Taylor series and higher order terms (cubic and higher) may be neglected to give

$$\phi_j\ (k) = 1 - \frac{k^2}{2n}\ \sigma^2$$

Defining $z = \Sigma_j\ y_j$ we obtain,

$$\phi_z\ (k) = \phi(k) = \prod_{j=1}^{n}\ \phi_j(k) = \prod_{j=1}^{n}\left(1 - \frac{k^2}{2n}\sigma^2\right) \approx \left(1 - \frac{k^2}{2n}\sigma^2\right)^n$$

$$\rightarrow \exp\left(-1/2\sigma^2 k^2\right) \equiv N(0, \sigma^2)$$

By transforming back to variable $\Sigma_j\ x_j$, we obtain the gaussian distribution for $\Sigma_j\ x_j$ as $N(\Sigma_j x_j, n\sigma^2)$

This can be extended for case where all $\sigma_j^2 \neq \sigma^2$.

Example 13

Generate a series of numbers distributed as pdf f(x) by Monte Carlo simulation.

Answer :

The sequence of operations to be followed is as follows :

(i) Generate r.n. x uniformly distributed within x_{min} to x_{max} i.e.

$$x = x_{min} + r_1(x_{max} - x_{min}) \text{ with } r_1 \text{ uniform over } [0, 1].$$

(ii) Generate a second independent random number u uniform over zero and f_{max} i.e. u = $r_2\, f_{max}$

(iii) if u < *f(x)*, then accept x. If not reject x and go to (i)

2.7 Exercises

1. Let r^{th} central moments be μ_r and $\beta_1 = \mu_3^2 / \mu_2^3$ and $\beta_2 = \mu_4 / \mu_2^2$. Then,

   prove that if $\begin{vmatrix} 1 & 0 & \mu_2 \\ 0 & \mu_2 & \mu_3 \\ \mu_2 & \mu_3 & \mu_4 \end{vmatrix} \geq 0$ then, $\beta_2 - \beta_1 - 1 \geq 0$.

2. Find M.G.F. and characteristic function of normal distribution with mean μ and variance σ^2.

   (*Ans.* $M(t) = e^{t\mu} + t^2 r^2/2$; C.F. = $M(it) = e^{it\mu} - t^2\sigma^2/2$)

3. If E is the expectation operator for the random variable X, i.e., $E(x^r) = \int x^r dF$, then prove that,

   (i) $E(cX) = cE(X)$ where c is a constant.

   (ii) $E(c_1X_1 + c_2X_2) = c_1E(X_1) + c_2E(X_2)$, c_i's are constants.

   (iii) $E(X_1\ X_2) = E(X_1)\ E(X_2)$ if X_1 and X_2 are independent.

4. Find the characteristic function of rectangular ($-a < x < a + \theta$) and Gamma densities.

$$\frac{\alpha^r}{\Gamma(r)} x^{r-1} e^{-\alpha x} \left(Ans.\, e^{it\theta} . \frac{e^{it\theta} - 1}{it\theta}, \left(\frac{\alpha}{(\alpha - it)} \right)^r \right)$$

5. If X, Y, Z are random variables, prove that following results on conditional expectation :
   (a) $Cov\,(X,Y) = Cov\,[X, E\,(Y|X)]$
   (b) $E(Z) = E\{E\,[V(Z|Y)|X]\}$
   $V(Z) = E\{E\,[V(Z|Y)|X]\} + E\{V\,[E(Z|Y)|X]\}$
   $+ V\{E\,[E\,(Z|Y)|X]\}$
   $Cov\,(X, Z) = Cov\,\{X, E\,[E(Z|Y)|X]\}$

6. Show that moments of two different densities as given below $[f_1(x)]$, $[f_2(x)]$ are same :
   $f_1(x) = (2p)^{-1/2}\, x^{-1} \exp\,[-(\log x)^2/2]$ *(log-normal)*
   $f_2(x) = f_1(x)\,[1 + a \sin(2\pi \log x)],\ -1 \leq a \leq 1$. *(not log-normal)*.

7. Find the density for the mean value $\bar{x}$ obtained from sampling 100 items from a unit normal population. (Ans. N (0, 1/10)).

8. Show that gamma distribution $f(x) = \dfrac{e^{-x/\beta} . x^{\alpha-1}}{\beta^\alpha \Gamma(\alpha)}$ has a single mode at $x = \beta(\alpha - 1)$ if $\alpha > 1$.

9. Find the error due to grouping frequencies at mid-point of class intervals in computation of cumulants. These are applicable to even cumulants only.

   (*Ans.* $(k_r)_c = k_r - c^r \dfrac{B_r}{r}, r \geq 2$. Where B_r is r^{th} Bernoulli number i.e. coefficient of $t^r/r!$ in expansion of $t(e^t - 1)^{-1}$. $(k_2)_c = k_2 - (c^2/12)$; $(k_4)_c = k_4 + (c^4/120)$; c is class intervals).

10. The Pareto distribution is useful for study of incomes and is defined as

$$f(x) = \frac{\alpha}{\beta}\left(\frac{\beta}{x}\right)^{\alpha+1}$$ $(x > \beta)$ and zero otherwise. Show that its rth moment exists only if $a > r$. If $a > 2$, find the first and second moments as well as its variance.

11. A sample of size N is taken from the exponential population with density e^{-x}, $x \geq 0$. Find the density function of its range and show that expectation of range is $1 + 1/2 + 1/3 + ... + 1/(N - 1)$.

    (*Hint* : Put $u = 1 - e^{-R}$ and expand $\log(1 - u)$ in series).

12. Numbers are drawn randomly from (0,1). How many are required before probability exceeds 0.95 such that range of sample is at least 0.5 ?

    (*Ans.* N given by $2^{N-2} > 5(N + 1)$).

References and Further Reading

1. Cramer, H., 1946,Mathematical methods of statistics: Princeton University Press.
2. Feller, W., 1965, An introduction to probability theory and its applications, Vol-I, 3rd Ed., New York, Wiley
3. Keeping, E. S., 1962, Introduction to statistical inference : Affiliated East West Press, New Delhi, 451p
4. Kendall, M. G. and Stuart, A., 1958etc., Advanced Theory of Statistics Vol.1 (3)., Griffin, London.
5. Pearson, E.S. and Hartley, H.O., 1954, Biometrika Tables for Statisticians, Vol.I, Cambridge University Press.
6. Rao, C. R., 1973, Linear statistical Inference and its Applications, New York, Wiley, 625p.

3 Estimation and Tests of Hypothesis

3.1 Introduction

Once population distribution function (d.f.) is identified (known), we have to estimate the parameters that are unknown. The parameters of a distribution can be estimated by use of characteristic function, moment generating function or cumulant generating function provided that these exist. It may be noted that characteristic functions always exist.

The cumulants (semi-invariants i.e., invariant except the first cumulant k_1) can be related to moments about origin (μ'_k) or central moments (μ_k) as follows:

$$k_1 = \mu = \mu'_1 = \mu$$

$$k_2 = \mu'_2 - \mu'^2_1 = \mu_2 \text{ (i.e., } - \sigma^2)$$

$$k_3 = \mu'_3 - \mu'_1 \mu'_2 + 2\mu'^3_1 = \mu_3$$

$$k_4 = \mu'_4 - 3\mu'^2_2 - 4\mu'_1 \mu'_3 + 12 \mu'^2_1 \mu'_2 - 6\mu'^4_1 = \mu_4 - 3\mu_2^2$$

or conversely,

$$\mu'_1 = k_1, \mu'_2 = k_2 + k_1^2, \mu'_3 = k_3 + 3k_2 k_1 + k_1^3, \mu'_4 = k_4 + 3 k_2^2 + 4 k_1 k_3 + 6 k_1^2 k_2 + k_1^4.$$

It is well known that if we have the distribution we can calculate its moments uniquely but the converse is not necessarily valid. A sufficient but not necessary condition for the distribution function (d.f.) to be uniquely defined by the moment sequence is that the series, $\sum_{i=1}^{\infty} \frac{\mu_i'}{i!} t$ is is absolutely convergent for some positive t. Any one of following sufficient conditions defines a unique d.f :

(a) The range of random variable is finite.

(b) $\sum_0^{\infty} \left(\mu_{2j}'\right)^{\frac{-1}{2j}} = \infty$, when range is ($-\infty$ to ∞).

(c) $\sum_0^{\infty} \left(\mu_j'\right)^{\frac{-1}{2j}} = \infty$, when range is (0 to ∞).

(d) $\lim_{j \to \infty} \sum \left(\mu_{2j}'\right)^{\frac{1}{2j}}$ is finite.

Example

After making probabilistic correction for observed moment sequence in thin section size analysis, we obtain a sequence of size moments in higher space (3*D* or 2*D* space) but have to compute the d.f.of grain size in the higher space (3D or 2*D*). This is done by using Edgeworth (Type *A*) expansion as range of size is finite (item a above). The desirable properties of estimators are discussed below :

Let T_n be an estimators of parameter θ based on first n samples estimators are used for estimating either parameter of distribution or the distribution function. The estimators should have following desirable properties.

Arithmatic mean $\overline{x_n}$ is a consistent estimators of mean (m) provided the sequence $\{x_n\}$ is iid and $E(x_i)$ exists (finite). For Cauchy distribution $\overline{x_n}$ is not consistent since $E(x_i)$ does not exist (infinite).

For normal population on which n samples are observed E (mle variance) $= \hat{\sigma}^2 = \frac{n-1}{n} \sigma^2$ which tends to σ^2 as $n \to \infty$, and its variance is $(2\,\sigma^4(n-1)/n^2) \to 0$ as $n \to \infty$. Hence mle estimators for variance σ^2 is consistent but biased.

Efficiency

An estimator T_n that attains minimum variance (lower bound) is called efficient. If it attains minimum variance (lower bound) as $n \rightarrow \infty$, then it is called asymptotically efficient. A consistent asymptotically normal (CAN) estimator for g (θ) has the least possible value of variance for $\sqrt{n}\,[T_n - g(\theta)]$ as $n \rightarrow \infty$.

Unbiasedness

T_n is unbiased (n finite or infinite) if $E(T_n) = \theta$. Whatever other parameters may occur in the distribution function. T_n is asymptotically unbiased if this happens only when $n \rightarrow \infty$.

Sufficiency

T_n is sufficient if no other estimator has greater information. Only arithmatic mean $\overline{x}$ is known to be sufficient statistic.

3.2 Maximum Likelihood (ML)

The concept is to choose and estimate θ through function of samplc obscrvation that will, when substituted for θ make the probability of sample is maximum. For example, for binomial distribution.

$$F(x \backslash \theta) = (N_x)\, \theta^x (1 - \theta)^{N-x}$$

This is maximum when $\partial f / \partial\theta = 0$ and $\partial^2 f / \partial^2\theta < 0$

Thus we obtain $\theta = x/N$

For a continuous variate $f(x|\theta, \theta_\alpha)$. The likelihood of sample set $x_1, x_2, \ldots x_N$ is $L = f(x_1|\, \theta, \theta_\alpha)\, f(x_2|\, \theta, \theta_\alpha) \ldots\ldots f(x_N|\, \theta, \theta_\alpha)$. This is always positive and L and $\log L$ have 1:1 relation, so maximizing $\log L$ (instead of L)

$\partial/\partial\theta\,(\log L) = 0$, $\partial^2/\partial^2\theta\,(\log L) < 0$ for normal proportion with mean μ, standard deviation σ

$$F(x \mid \mu, \sigma) = (2\pi\sigma^2)^{1/2} \exp(-(x - \mu)^2/\sigma^2)$$

And $$\text{Log } L = -N/2 \log 2\pi - N \log \sigma - \sum_{i=1}^{N} (xi - \mu)^2 / \sigma^2$$

Hence $$\partial (\log L)/\partial\mu = \Sigma_i (xi - \mu)/\sigma^2$$

And $$\partial^2 (\log L)/\partial^2\mu = -N/\sigma^2$$

The maximum likelihood $\hat{\mu}$ is therefore $\Sigma_i (x_i - \mu) = 0$

or $$\hat{\mu} = 1/N\ \Sigma x_i = \overline{x} = \text{m (sample mean)}.$$

This value is independent of other parameter σ.

Similarly, $$\partial (\log L)/\partial\mu = -N/\sigma + \Sigma_i (x_i - \mu)^2/\sigma^3$$

Equating above to zero we see that the value of $\hat{\sigma}$ is not independent of μ. There is no MLE of σ itself but we can jointly solve $\partial (\log L)/\partial\mu = 0$ and $\partial (\log L)/\partial\sigma = 0$.

This gives $\hat{\mu} = m$ and $\hat{\sigma}^2 = 1/N \sum_i (x_i - m)^2$

However, *m* is unbiased estimate of μ but σ^2 (sample variance) is biased. It can be corrected to $\hat{\sigma}^2$ (unbiased) $= N\,\hat{\sigma}^2/(N-1)$.

ML estimators are the best available statistics. They are consistent, tend to be normally distributed as *N* increases, most efficient, sufficient as well as they are invariant under functional transformation.

For a Cauchy distribution median *u* may be better than sample mean $\overline{x}$.

$$\text{cauchy pdf} = f(x) = (1/\pi)\ (1/\pi + (x-\theta)^2) \text{ has}$$

the sample median u has a value $\pi^2/4N$.

Its MLE is

$$\partial (\log L)/\partial \mu = 2 \sum_i (xi-\theta)/1 + (xi-\theta)^2 = 0$$

$$-[v\,(T)]^{-1} = 2N/\pi \int_{-\infty}^{\infty} (x-\theta)^2 - 1\ dx/[(1+ (x-\theta)^2]3$$

$$\doteq 4N/\pi \int_{-\infty}^{\infty} (u^2-1)\ du/(1+u^2)^3$$

$$= -N/2.$$

$$\text{or, } v(T) = 2/N$$

Often we require standard error of estimates so that the confidence interval obtained for point estimates. ML estimator can be biased, but such bias corrections are easily made.

Example : ML estimate of normal distribution

Let *N* samples were observed for a population, so that the log likelihood *L* is given by

$$\ln L = -(N/2) \ln (2\pi) - N \ln \sigma - \sum (x-\mu)^2/2\,\sigma^2$$

Differentiating with respect to parameters and equating to zero, we get

$$\partial (\ln L)/\partial\mu = \sum (x-\mu)^2/2\,\sigma^2 = 0$$

$$\partial (\ln L)/\partial\sigma = -(N/\sigma) + \sum (x-\mu)^2/2\,\sigma^3 = 0$$

Thus the ML estimates are $\hat{\mu} = 1/N\ \sum x_i = \overline{x} = m$

But the second equation has a nuisance parameter μ, so we can not get σ. But joint MLE for μ and σ can be obtained by solving above two equations together with substituting *m* for μ in second equation. Thus we get $\mu = \overline{x} = m$, $\sigma^2 = (1/N)\ (x-m)^2 = m_2$ (sample second moment).

We know, $\overline{x}$ = m is unbiased but m_2 is biased as $E(m_2) = (N-1)\,\sigma^2/N$.

We can get unbiased estimator of population variance as :

$$\sigma_u^2 = (N/(N-1))m^2 = \Sigma\,(x-m)^2/(N-1)$$

Properties of Maximum Likelihood Estimators

The following five properties are main reasons for recommending the use of maximum likelihood (ML) estimators :

(a) The ML estimator is *consistent.* If $f(x \mid \theta)$ is continuous in x, and also continuous and monotonic in θ over an interval including the true value θ_0, and if T is the ML estimator of θ, then T converges in probability to θ_0 as the sample size increases. The proof holds also for a discrete variate if we replace each value by an interval over which we suppose that the frequency distributed uniformly.

(b) The ML estimator tends to *normality* as N increases. The conditions in (a) are supposed to hold, together with some further conditions on the continuity of $\partial f / \partial\theta$.

(c) The ML estimator is *most efficient.* The variance of the ML estimator T is given by

$$[V(T)]^{-1} = N\int_{-\infty}^{\infty}\left(\frac{\partial \log f}{\partial\theta}\right)^2_{\theta=\theta_0} fdx = NE\left[\left(\frac{\partial \log f}{\partial\theta}\right)^2\right]_{\theta=\theta_0}$$

where, on the RHS, θ is to be put equal to θ_0. If the domain of f does not depend on θ, above equation is equivalent to

$$[V(T)]^{-1} = -NE\left[\left(\frac{\partial^2 \log f}{\partial\theta^2}\right)\right]_{\theta=\theta_0}$$

Since log $L = \Sigma_i \log f(x_i)$, and since the x_i all have the same distribution, second equation may be written

$$[V(T)]^{-1} = -E\left[\left(\frac{\partial^2 \log f}{\partial\theta^2}\right)\right]_{\theta=\theta_0}$$

This is a convenient way of finding the variance of a ML estimator. It can be shown that if U is any estimator for θ, which in large samples is normally distributed about θ_0 with variance $V(U)$, and if the domain of f does not depend on θ, then

$$[V(T)]^{-1} \leq NE\left[\left(\frac{\partial^2 \log f}{\partial\theta}\right)^2\right]_{\theta_0}$$

It follows from first equation and property (b) above that the ML estimator is most efficient.

(d) If a sufficient estimator exists for θ, the ML estimator is *sufficient.*

If T is sufficient and $g(t \mid \theta)$ is its density function,

$$L = g(t \mid \theta) \, . \, h(x_1, x_2, \ldots\ldots x_N \mid t)$$

where h is a function of the sample values which does not depend on θ.

Therefore, $\dfrac{\partial}{\partial\theta}(\log L) = \dfrac{1}{g}\dfrac{\partial g}{\partial\theta} = \psi(\theta, t)$

The ML estimator is given by putting $\psi(\theta, t) = 0$ and solving for θ. The result is obviously a function of t, say $\phi(t)$. The estimator is therefore, $\phi(T)$, and since T is sufficient, so is $\phi(T)$.

(e) The ML estimator is *invariant* under functional transformations. This means that if T is the ML estimator of θ, and if $u(\theta)$ is a function of θ, then $u(T)$ is the ML estimator for $u(\theta)$.

If for a example, we are dealing with a normal population (for which $m_4 \doteq 3\sigma^4$) and we know that the ML estimator for σ^2 is the sample second moment m_2 (i.e., $N^{-1} \sum (x_i - m)^2$), we conclude that the ML estimator for m_4 is $3m_2^2$ and not m_4. Of course, m_4 might be used as an estimator, but it would not have as small sampling variance as $3m_2^2$.

This property is not true of all estimators. If T is an unbiased estimator of θ, for example, it does not follow that T^2 is an unbiased estimator of θ^2.

The Cramer - Rao Inequality

If T is an estimator of θ, and if its *bias* is b, as defined by the relation

$$E(T) = \theta + b \qquad \ldots\ldots(3.1)$$

where b may depend upon θ, then

$$E\left[(T-\theta)^2\right] \geq \frac{\left(1 + \dfrac{db}{d\theta}\right)^2}{N\int_{-\infty}^{\infty}\left(\dfrac{\partial \log f}{\partial\theta}\right)^2 f(x \mid \theta)\,dx} \qquad \ldots\ldots(3.2)$$

If $g(t \mid \theta)$ is the frequency function for T, the equality sign in relation (above mentioned) holds if, and only if,

(a) T is sufficient (b) $\frac{\partial \log g}{\partial \theta} = k(T - \theta)$ .....(3.3)

where k is independent of T but may depend on θ.

If T is an *unbiased* estimator, so that $b = 0$, $E[(T - \theta)^2]$ becomes that variance of T, and so eqn. (2) is

$$V(T) \geq \left[N \int_{-\infty}^{\infty} \left(\frac{\partial \log f}{\partial \theta} \right)^2 f dx \right]^{-1} \quad .....(3.4)$$

which is formally the same as (3), although the conditions imposed on the estimator are different. The relation (4) was proved independently by Cramer and Rao, although it was found earlier by Fisher for the special case of a normal population.

3.3 Parametric Tests of Hypothesis

Often we wish to test hypothesis whether a sample belongs to a population (characterised by its parameters) or whether two or more samples are homogeneous. Such hypothesis is called null hypothesis (H_0) and sample statistics obtained by experiments are tested against a suitable alternative hypothesis (H_1). Based on the sample following decisions are possible:

(i) reject null hypothesis

(ii) accept null hypothesis

(iii) more experimentation (samples) are required to take any decision. (sequential sampling until decision i or ii is taken).

These actions involve the risk of taking any decision as two kinds of errors could be dangerous. Rejection error (error of first kind or α error) we reject H_0 when it is true. We may accept H_1 when false (acceptance error or error of second kind or β error), which is most dangerous.

We compute test statistics much as arithmetic mean or variance from sample data and check if these computed value lies within time rejective interval or sets of intervals along real axis. In some are cases the hypothesis may include a pair of numbers, so that the rejection refine is apart of two-dimensional plane.

Any hypothesis that completely specifies the distribution is called *simple* (Fig. 3.1), otherwise it is called *composite* (Fig. 3.2 & 3.3). In case of normal population with two parameters mean μ and variance σ^2, suppose σ^2 is given and we wish to test H_0 = (simple) but if we wish to test $H_0 = \mu_0$ or $H_1 = \mu_1$ ($\neq \mu_0$) then alternative H_1 is a composite hypothesis. The alternative $H_1 = \mu_1$ could be two sided ($\mu_1 > \mu_0$ and $\mu_1 < \mu_0$) or one sided ($\mu_1 > \mu_0$ or $\mu_1 < \mu_0$) (Fig. 3.4(a, b)).

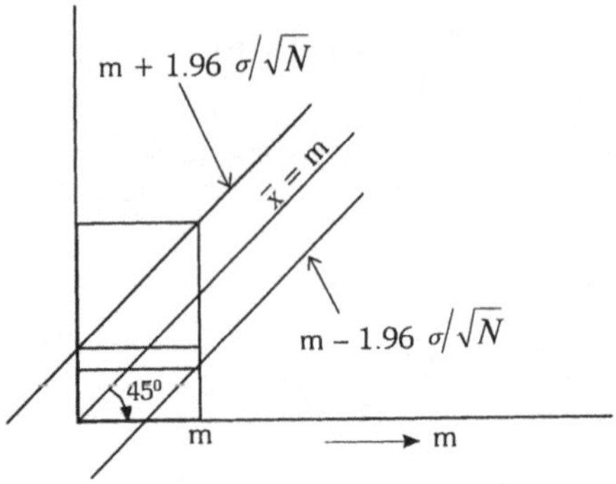

Fig. 3.1 95% confidence interval for mean in a Gaussian p.d.f. with known σ^2 (z-test)

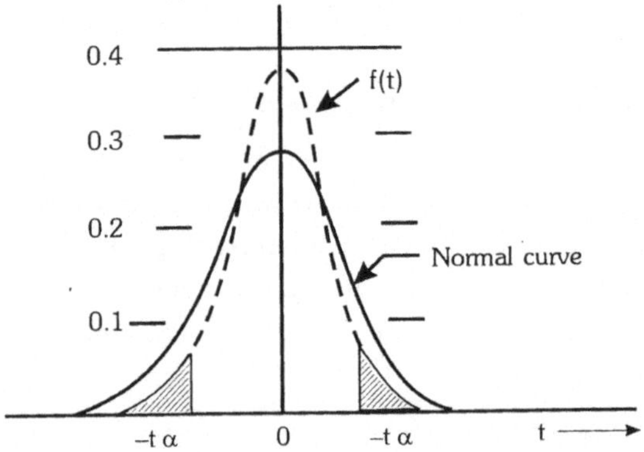

Fig. 3.2 Graphs for p.d.f.s for *t* and Normal distributions

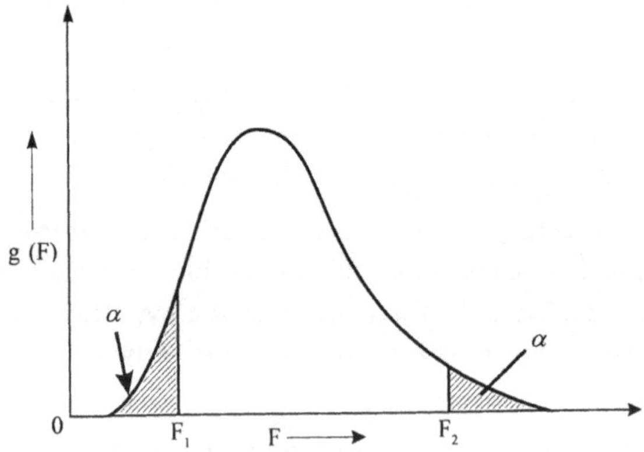

Fig. 3.3 F - distribution

The test probability under $H_0 = \theta_0$ for a simple hypothesis for a test statistics is the size of the test a error is given by the integral $\alpha = \int_R g(t/\theta_0)\,dt$. In practice α is chosen as 0.05,0.01 or 0.001, (levels of probability).

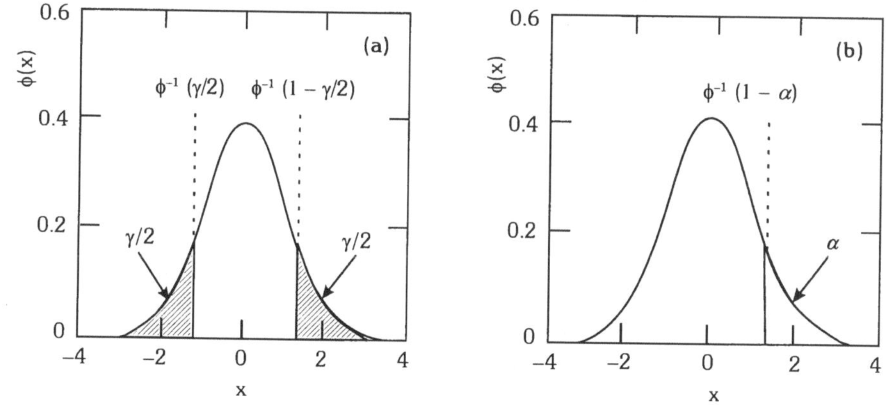

Fig. 3.4 (a) Standard Gaussian p.d.f. showing relationship between quantities ϕ^{-1} and Central Confidence interval
(b) Standard Gaussian p.d.f. showing relationship between quantities ϕ^{-1} and one sided Confidence interval

Similarly if H_1 is true we comment on acceptance error (β) as $\beta = \int_{(A)} g\,(t/\theta_1)dt$, where A is region of acceptance. The power of the test is $P(\theta_1) = 1 - \beta = \int_{(A)} g\,(t/\theta_1)dt$. The test should be as powerful as possible for the same size (Fig. 3.5).

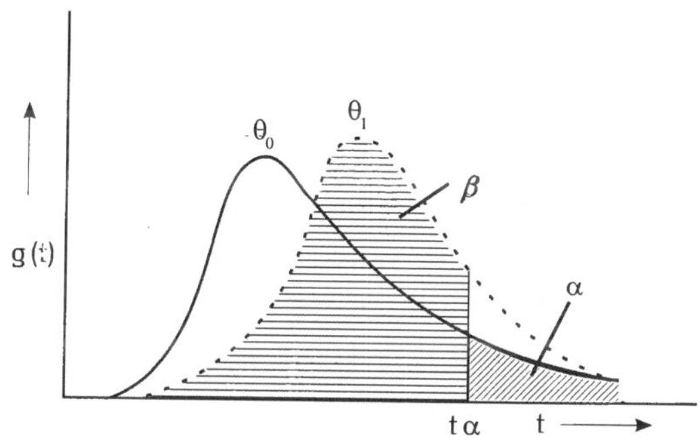

Fig. 3.5 Errors of first (α) and second kind (β)

If θ is of any value of θ_1 the $P(\theta)$ is called power function of test for θ_0 is equals θ. If θ is near θ_0, the power will be small and if $\theta = \theta_0$, then $P(\theta) = \alpha$. If θ is far away from θ_0, $P(\theta)$ tends to 1. In industry we use operating characteristic (o.c.) which is $\beta\,(\theta) = 1 - P\,(\theta)$ (Fig. 3.6).

A test using R' (lying above R for $\theta_1 > \theta_0$) may be more powerful but it is less powerful is R' lies below R for $\theta_1 < \theta_0$. If every value of θ (except θ_0), the power curve of R' is below R, the test is uniformly more powerful than using R'. If this holds the all-possible devices of R' then the test using R is a uniformly most powerful test (Fig. 3.7).

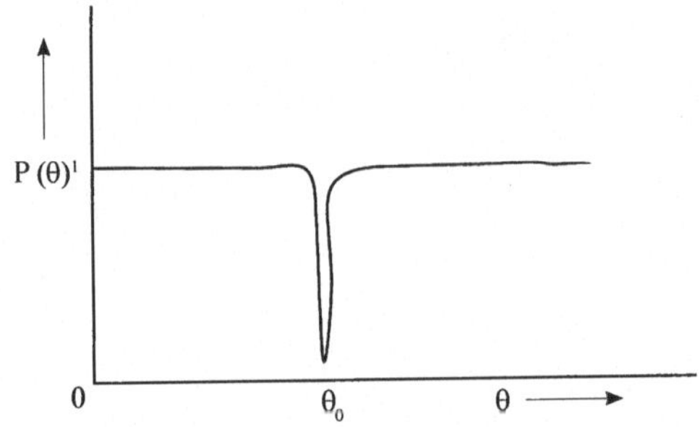

Fig. 3.6 Power function of an ideal test with alternatives having non-zero probabilities

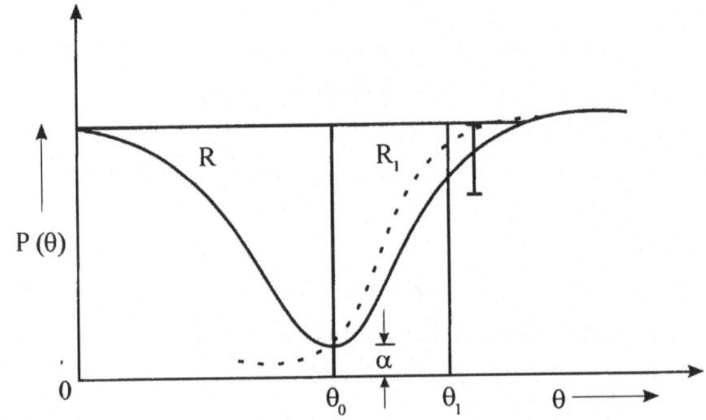

Fig. 3.7 Power functions of alternative tests

According to Neyman & Pearson (1933), if exists for a region, $\alpha = \int_{(R)} f(x/\theta_0)dx$ and such that $x \in R$ whenever the likelihood ratio $L(x) = (f(x/\theta_0)/f(x/\theta_1)) < c$ (constant) and does not belong to R whenever this ratio is $> c$ then R is a most powerful test of size $\leq \alpha$. When N is large the distribution of $-2 \log L(x)$ under H_0 is a chi square distribution with degrees of freedom depending on the parameter concerned and it is distributed as non-central chi square if H_1 is true.

We can increase the power of a test by calling a randomized decision. The dimension of the test statistics X is partitioned in to three parts : R, A and D. If X falls in R, H_0 is rejected, if x falls in A, H_0 is accepted and if it falls in D we toss a coin or use random numbers to decide the known probability of rejection of H_0.

We have

$$\psi(x) = 1 \text{ if } L(x) < c$$

$$\psi(x) = 0 \text{ if } L(x) > c$$

$$\psi(x) = \psi_0 \text{ if } L(x) = c$$

The test function $\psi(x)$ is most powerful of size α for testing H_0 against H_1. The value of c is generally $P(L(x) < c \mid H_0) \leq \alpha$ and value of ψ_0 by $P(L(x) < c \mid H_0) + \psi_0 P(L(x) = c \mid H_0) = \alpha$.

Bayes' Rule

$$P(\theta \in \omega/x) = \frac{\int_{(\omega)} P(x/\theta) p_\theta \, d\theta}{\int_{(\Omega)} p(x/\theta) p_\theta \, d\theta}$$

where p_θ = prior probabilities.

However, prior probabilities are not always available (known). To avoid estimation of prior probabilities for Bayes' principle, Wald suggested a procedure to minimize the maximum risk. But Wald's criterion may be unduly pessimistic and approximation using Game Theory would be better.

Example :

In a betting situation one wins α and losses β and probability of success is θ. The expected loss per bet is $\beta(1 - \theta) + \alpha\theta$ which is positive if $\theta < \theta_0$ where $\theta_0 = \beta/(\alpha + \beta)$. The risk function assuming each bet is independent binomial is

$$\gamma(n)\theta = [\beta(1 - \theta) - \alpha\theta] \, . \, B(n, N, \theta), \theta \leq \theta_0$$

$$\gamma(n)\theta = [\alpha\theta - \beta(1 - \theta)] \, . \, [1 - B(n, N, \theta)], \theta \geq \theta_0$$

B is approximately $1 - \phi(z)$ with $z = (n - (1/2) - N\theta)/[N\theta(1 - \theta)]^{1/2}$

The minimum risk occurs when maximum of γ for $\theta \leq \theta_0$ is equal to the maximum of γ for $\theta \geq \theta_0$. This maxima occurs at $\theta \approx \theta_0 + 0.752 [\theta_0 (1 - \theta)/N]^{1/2}$. We take n as $N\theta_0 = N\beta/(\alpha + \beta)$ and to bet the next time if $x \geq n$. This is Wald's minimax principle.

3.4 Non-Parametric Tests for Hypothesis

Tests which do not require assumptions about the parameters of the population are called non-parametric (distribution free) tests. The samples drawn from the population are never assumed to be independent of each other. Whenever normalizing assumption can not be made for the population (z, t, F tests could be dangerous) non parametric tests are useful and robust.

(a) *χ^2 test of Goodness of fit :*

$\chi^2 = \Sigma((O - E)^2/E)$ with suitable degrees of freedom for χ^2 test.

(b) *Kolmogorov*

Observed cumulative frequency = $S_N(x) = k/N$

Length upper D_N = lub $|S_N(x) - F(x)|$

Kolmogorov proved that $\lim_{N \to \infty} Pr(N^{1/2} D_N \geq \lambda) = L(\lambda)$

where $L(\lambda) - 2 \Sigma_{n-1}^{\infty} (\ 1)^{n+1} e^{-2n?} \lambda^2$

Thus, for $\lambda = 1.36$, $L(\lambda) = 0.05$ giving the asymptotic probability of D_N exceed $1.36 \, N^{1/2}$ take a value of 0.05.

(c) *K - Smirnoff test*

D_{mn} = lub $|Sm(x) - Sn(x)|$ and $\text{Lim}_{mn \to \infty} P(N^{1/2} D_{mn} \geq \lambda) = L(\lambda)$ (as above) and $N = mn/(m+n)$. We assume $mn \to \infty$ such that m/n is finite. For large m,n, values for λ for 0.05 and 0.01 levels are 1.36 and 1.63 respectively.

(d) *Tests of randomness*

If sample data falls outside $x \pm 3\sigma$ then these data are either outliers and not belonging to the same population or samples are not independent.

(e) *Runs up and down*

The probability of exactly k positive difference are $\phi_N(k)/N!$.

Then $$\phi_N(k) = \sum_{i=o}^{k}(-1)^i \binom{N+1}{i}(k+1-i)^N$$

The expectation is $E(k) = \sum_{k=0}^{N-1} \frac{k\phi_N(k)}{N!} = \frac{N-1}{2}$ and variance is

$$V(k) = (N+1)/12.$$

For $N > 12$, we can use normal approximation (with continuity correction) and $z = \pm\left(3/(N+1)^{1/2}\left(|2k - N + 1|\right) - 1\right)$.

3.5 Solved Examples

Example 1 :

Optimal sample size for estimating mean with a fixed error and confidence level. How large a sample should be collected to estimate mean (at an error not exceeding 0.5) of a Normal population at confidence level 0.8, with variance 9 ?

Answer :

We have $X_i \sim N(\mu, 9)$. So $\overline{X}_n \sim N((\mu, 9/n)$. We want $P\left(|\overline{X}_n - \mu| < 0.5\right) > 0.8$.

The standard normal deviation

$$z = \sqrt{n}\left(\overline{X}_n - \mu\right)/3$$

and we have $P\left(|z| < \frac{0.5\sqrt{n}}{3}\right) => 0.8$ which yields $n > 59.13$.

Therefore, we need sample size of 60.

Example 2 :

An urn has red and white balls with probability of red balls known to be $p = 1/3$ or $p = 1/2$. We draw five balls with replacement and decide whether $p = 1/3$ or $1/2$.

Answer :

The Bayesian prior belief is $P\ (p = 1/3) = 3/4$, $P\ (p = 1/2) = 1/4$.

The observation was 3 red and 2 white balls.

The posterior probability A as calculated by Bayes' Theorem is as follows :

$$P\left(P=1/3 \mid A\right) = \frac{P\left(A/\ p=1/3\right)P\left(p=1/3\right)}{P\left(A/p=1/3\right)P\left(p=1/3\right)+P\left(A/p=1/2\right)P\left(p=1/2\right)}$$

$$= \frac{{}^5C_3\left(1/3\right)^3\ \left(2/3\right)^2\ \left(3/4\right)}{{}^5C_3(1/3)^3(2/3)^2(3/4)+{}^5C_3(1/2)^3(1/2)^2(1/4)} \cong 0.61$$

$(p = 1/2 \mid A) \cong 0.39$ which is higher than prior belief (= 1/4).

However, Bayesian inference based on uniform prior density leads to results similar to classical statistical inference.

Example 3 :

Point estimation of parameter θ, given posterior density $f_1(\theta)$, for $B(n, p)$

Answer :

If $\hat{\theta}$ is the estimate, assume loss function as, Loss = $(\hat{\theta} - \theta)^2$

Bayesian chosen $\hat{\theta}$ to minimize $E\left(\theta-\hat{\theta}\right)^2 = \int_{-\infty}^{\infty}\left(\theta-\hat{\theta}\right)^2 f_1(\theta)\ d\theta$ as $\hat{\theta}$ is fixed

and θ is random variable. Equation $dE/d\theta = 0$. We get, $2\int_{-\infty}^{\infty}\left(\hat{\theta}-\theta\right) f_1(\theta)\ d\theta = 0$

or $\hat{0} = \int_{-\infty}^{\infty}(0)\ f_1(\theta)\ d\theta$.

$\hat{\theta}$ is Bayes' estimator under squared error loss. When n is small, Bayesian point estimation $\left(\hat{p}_{Bayes} = \frac{k+1}{n+2}\right)$ has a better efficiency that ML estimator $\left(\hat{p}_{ML} = \frac{k}{n}\right)$ for $B(n, p)$ where, p = success probability.

Example 4 :

Advantages (* items in the Table) of Bayesian and classical schools of statistics.

Answer :

Bayesian	Classical
1. Can make exact interference using posterior distribution.*	1. Use confidence intervals as subtitute
2. Bayes estimator is good, even by classical standards.*	2. If sample size is large, MLE is just as good.
3. Bayes inference may be robust against mis-specifier of distribution.	3. Can use good estimator such as sample mean without assuming any distribution.*
4. Use prior distribution that represents total ignorance.	4. Objective inference.*
5. No need to obtain distribution of estimater.*	5. No need to calculate complicated integrals.*

Example 5 Test of Hypothesis

A Simple hypothesis specifies values of all the parameters of a distribution, otherwise it is called Composite.

Answer :

$H_0 : p = 1/2$ is a simple hypothesis but $H_1 = p > 1/2$ is a composite hypothesis for Binomial distribution. Different hypothesis testing possibilities are :

(i) Simple vs simple (use $\alpha = \beta$);

(ii) Simple vs composite (use power function β) and

(iii) Composite vs composite (use Likelihood Ratio (LR) test which has good asymptotic property).

In addition to the above parametric tests of hypothesis we can also use a number of Non- Parametric tests.Let X have density $f(x) = 1/\theta$ for $0 \leq x \leq \theta$ and zero otherwise. $H_0 : \theta = 1$ against $H_1 : \theta > 1$, $N = 1$ on X. Find power function of test based on critical region $R = [0.75, \infty]$.Power function = $P(X \geq .75 \mid \theta)$ $= 1 - 0.75/\theta$, which is 0.25 at $\theta = 1$ and exponentially grows to 1 as θ tends to ∞. *Uniformly Most Powerful* (UMP) test has maximum power against other tests. However, UMP may not exist if H_0 is simple and H_1 is composite. In such cases, likelihood ratio (LR test) is used which has good asymptotic properties. For example, $-2 \log l$ has asymptotic χ^2-distribution with distribution function equal to the number of restrictions implied by H_0.

3.6 Case Studies

Roundness Measure Ranging from (0 to 1)

Often concepts are clearly defined but operational procedures for measurement may not be truly reflecting the concepts. This results in inconsistent measures such as Wadell's roundness.

Roundness of particles in $2D$ is defined as varying from zero (perfectly angular) to unity (circle). As grains are of varying shapes and roundness, some confusion of measures are existing in literature. Wadell suggested to define roundness as the ratio of average radius of curvature of corners to radius of inscribed circle (R_i) and he suggested that if a radius of corner is greater than inscribed circle it be replaced by R_i (otherwise values will be more than 1 hence inconsistent).

However, the above Wadell procedure results in roundness of elliptical outlines to vary from 0.5 to 1 since ellipse has four corners with curvatures b^2/a and a^2/b for two opposite corners ($Ri = b$). For, rectangular (prismatic) particles Wadell procedure correctly give roundness ranging from zero to unity.

Sahu (1976) modified Wadell procedure so that ellipse has two corners only with radii of curvature b^2/a (all corners whose radii of curvature are greater than R_i should not be called corners and dropped from measurements). This yielded roundness varying from zero to unity, which is consistent, and as desired.

Independence of Sphericity and Roundness

In sedimentology, (conceptually) sphericity and roundness of particles are independent. Sphericity is defined for $3D$ as ratio of surface area of sphere of equivalent volume to that of particle and range from 0 to 1. In $2D$, sphericity can be perimeter of a circle of equivalent area to actual perimeter or by ratio of smaller to greater axis. Often $3D$ sphericity is measured in sections or projections in $2D$ for convenience.

Considering ellipse, sphericity can be defined as $\psi_e = b/a$ whereas roundness ρ_e is given by

$$\rho_e = \frac{\left(\frac{b^2}{a} + \frac{b^2}{a}\right)}{2b} = \frac{b}{a}$$

$$= \psi_{e.}$$

Thus, for ellipse, sphericity and roundness, are equivalent and strongly correlated ($\rho = 1.0$). For rectangular outlines, sphericity can be defined as $\psi_t = b/a$ (as for ellipse) but $\rho_r = \sum_{i=1}^{4} \frac{r_i}{4b}$, which can range from zero to 1.Hence, ρ_r is independent of ψ_r. If a sample has α proportion of elliptical and $(1-\alpha)$ proportion of rectangular outlines, then, $\left(\frac{\bar{b}}{a}\right) = \overline{\psi} = \alpha\ \overline{\psi}_e + (1-\alpha)\ \overline{\psi}_r = \overline{\psi}$ as $\left(\overline{\Psi}_e \cong \overline{\Psi}_r\right)$.

But, $\bar{\rho} = \alpha\, \bar{\rho}_e + (1-\alpha)\bar{\rho}_r$

or $\bar{\rho} = \alpha\, \bar{\Psi}_e (1-\alpha)\, \bar{\rho}_r = \alpha\, \bar{\Psi} + (1-\alpha)\, \bar{\rho}_r$

which shows linear regression between $\bar{\rho}$ and $\bar{\Psi}$ with intercept depending on $(1-\alpha)\, \bar{\rho}_r$.

Use of Partial Correlation

Ratha (2000) has studied multiple (no asterisk) and partial correlation (asterisk) among several toxic elements in rocks and soils around Mumbai in order to focus attention to real correlation among these elements (without interference of other toxic elements).

The results for 35 samples are listed as follows :

	Basaltic Rocks				Residual Soils			
	Cd	Cu	Pb	Zn	Cd	Cu	Pb	Zn
Cu	.08	1.0			.02	1.0		
	–.04				–.25			
Pb	–.21	.19	1.0		.38	.10	1.0	
	–.07	–.29			.05	.26		
Zn	.04	.46	.61	1.0	.23	–.15	.06	1.0
	–.31	.37	.60		.01	–.09	.02	

In soils, the only multiple correlation which is significant was Cd-Pb (.38) but none of the partial correlations are significant. Hence, it is inferred that natural weathering and leaching processes have removed all toxic elements from the residual soils around Mumbai. However, in rocks (basalts) Cd-Zn has significant negative partial correlation, which may be due to lattice substitution in minerals. But Cu-Zn and Pb-Zn are significantly positively partially correlated which indicate that these pairs do not substitute each other in the lattice of minerals of these basalts.

3.7 Exercises

1. Show that $\bar{x}_n = \sum_{i=1}^{n} x_1/_n$ is an efficient unbiased estimator for parameter θ in a Poisson distribution given by $e^{-\theta}\, \theta^x / x\,!$, where $x = 0, 1, \ldots$..

2. Calculate the mean and variance of $z=\sum_{i=1}^{n}(X_i-\mu)^2/n$ for parameter θ in $f(x \mid \theta) = N(m, \theta)$. Calculate the efficient unbiased estimator for θ.
3. Find the MLE of θ for $f(x \mid \theta) = \theta e^{-\theta x}$, $x > 0$, based on a random sample of size n. What is the MLE of EX. Compare these two results.
4. Use method of moments to estimate parameter θ for density $f(x \mid \theta) = x\exp[-(x^2/2\theta)]/\theta$ where $x > 0$, $\theta > 0$, based on random sample of size n.
5. If $\bar{x} = 20$ for sample size $n = 25$ from a normal population $N(\mu, 16)$, find a 90% confidence interval for population mean μ.
6. Find Bayes' estimator for Poisson parameter μ based on a random sample of size n and quadratic loss = $(a - \mu)^2$ and $f(\mu) = e^{-\mu}$ with $\mu > 0$. Loss function could be quadratic, absolute difference (robust) $|\alpha - m|$ or power of absolute difference, i.e., $c(\theta)\,|a - \mu|^{\phi}$.
7. Estimate $E(1/\sqrt{x})$ based on a random sample of size n when density X is not known. Compare if X has an exponential density $e^{-(x/\theta)}/\theta$, $x > 0$.

References and Further Reading

1. Cramer, H., 1946, Mathematical Methods of Statistics, Princeton University Press
2. Fisher, R. A.,1956, Statistical Methods and Scientific Inference, Edinburgh, Oliver and Boyd
3. Keeping, E. S., 1962, Introduction to Statistical Inference, Affiliated East-West Press, NewDelhi, p451
4. Luce, R. D. and Raiffa,H., 1957, Games and Decisions, NewYork, Wiley
5. Neyman, J. and Pearson, E. S., 1933, On the problem of the statistical hypothesis, Phil. Trans.Roy. Soc. A231:p289-337
6. Rao, C. R., 1973, Linear Statistical Inference and its Applications, NewYork, Wiley, p625
7. Ratha,D.S., 2000, Applications of Statistical Techniques in Environmental Geochemistry
   In : Geological Data Analysis :Statistical Methods ,Hindustan Publishing Company, NewDelhi
8. Sahu, B.K., 1976, Mathematical Theory of counting two dimensional grains by line and ribbon methods : Sedimentary Geology, 16:177-92.

4 Linear Statistical Models

4.1 Introduction

Size and shapes of rocks and fossils and pieces of earth materials cannot be defined through a single number (scalar) but through a set of vector numbers in a grid space. Vectors belonging to a field follows commutative and associative laws of addition implying existence of zero element (like zero in number system). Vectors also follow commutative, associative and distributive laws implying existence of unit element (like unity in number system). A collection of elements satisfying all these laws forms a linear space (finite or infinite).

Vector addition is defined as $x + y = y + x$ (commutative), $x + (y + z) = (x + y) + z$, along with a null vector 0 such that $x + 0 = x$ for all x. Inverse vector (ξ) exists such that $x + \xi = 0$.

Scalar (c) multiplication of vector leads to cx and is subjected to commutative and distributive laws for vectors and scalars, associative law for scalars and existence of unit scalar c such that $cx = x$.

A set of vectors u_1 ..... u_k is linearly depends if there exists scalars c_1 ..... c_k not all zero such that $c_1u_1 + ... + c_k u_k = 0$, otherwise it is independent. Here the null vector is a dependent set, a set of vectors containing null vector is dependent and a set of non-zero vectors $u_1 \,.....\, u_k$ is dependent when and only when a member of this set is linear combination of predecessors. A linearly independent subset of vectors in the vector space, generating or spanning it is called a basis of the linear vector space. Every vector space has a basis. If $\alpha_1 \,.....\, \alpha_k$ and $\beta_1 \,.....\, \beta_s$ are two alternative choices for basis, then $s = k$ (dimension/rank). Every vector in the vector space has a unique representation in terms of a given basis. Any two finite dimensional vector spaces are isomorphic i.e., if

$$u \to u^*, v \to v^* \text{ then } u + v \to u^* + v^* \text{ and } cu \to cu^*$$

Consider linear equation $x_1\,\alpha_1 + \ldots + x_m\,\alpha_m = 0$. It has numerical solution (not all x_i zero simultaneously) where α_i's are dependent. This solution (row vectors in *Em*) constitute a vector space. The row linear equation $x_1\,a_1 + \,.....\, + x_m\,\alpha_m = \alpha_0$ has solution if α_0 is dependent on $\alpha_1 \ldots \alpha_k$. A set of solutions belong to *Em* but do not constitute a subspace as in the case of homogeneous equations. The general solution of such equation is sum of a particular solution and solution for homogeneous equation.

Inner Product

In addition to independence (dependence) of elements of vector spaces, we should consider distance and angle between the vectors. When the associated scalars belong to real or complex numbers, we extend these concepts through a complex valued function called inner product of two vectors (x,y) which satisfy the following conditions :

$$(x,y) = \left(\overline{y,x}\right), \text{ complex conjugate of } (y, x)$$

$$(x, x) > 0 \text{ if } x \neq 0, = 0 \text{ if } x = 0$$

$$(\phi\, x, y) = \phi(x,y), \text{ where } \phi \text{ is a scalar}$$

$$(x + y, z) = (x, z) + (y, z)$$

For R^n, *n*-dimensional enclosed space with real coordinates, the real value function $(x, y) = \sum x_i\; y_i$ satisfies the above four conditions.

For U^n, unit space with complex coordinates, the inner product is of the function $\sum a_{ij}\, x_i\; y_i$, where a_{ij} are chosen suitably. The positive square root of (x, x) depends on $\|x\|$ – norm of *x* or length of *x*. The function $\|x - y\|$ satisfies the postulates of a distance on a metric space(i.e. distance between *x* and *y*).

The triangle inequality is given by $\|x - y\| + \|y - z\| \geq \|x - z\|$ and Cauchy-Schwartz inequality by $|(x, y)| \leq \|x\|\;\|y\|$. The vectors are independent if their inner product is zero. i.e. $(x,y) = 0$, and angle θ between non null vector is defined as $\cos\theta = (x, y) \,/\, \|x\| \cdot \|y\|$ with range $[-1, +1]$.

In particular, linearity helps bring simplicity in mathematical analysis as well as statistical inference. For example, linear combinations of normally distributed random variables have normal distributions. This is also true for exponential and gamma densities.

4.2 Analysis of Variance (ANOVA) Models

ANOVA is widely used for identifying separate sources of variance in a set of samples. The samples may have suffered different treatments, which may induce different mean levels for the random variable. If so, limited samples in one grand sample would have a variance partially due to variances within each treatment and variance due to the effect of different treatments. This enables us to decide whether any treatment has produced significant effects. Some standard experimental designs will be discussed here.

All ANOVA tests are based on following assumptions :

(i) Normality of random variable X

(ii) (Linearity) additive treatment effects

(iii) Constancy of variance between the samples

The homogeneity of variances for two samples with $n1$ and $n2$ degrees of freedom for their variances s_1^2 & s_2^2 respectively can easily is tested by F-ratio test. Therefore, we consider k samples ($k > 2$) where i^{th} sample of size N_i and observed values are x_{ij}, $i = 1, 2, \ldots k$, $j = 1, 2, \ldots\ldots N_i$. Assuming as required all samples are independent and come from normal populations with means μ_1 ..... μ_k and variances σ_1^2 ..... σ_k^2, the null hypothesis (Ho) is all k variances are equal and the alternative hypothesis ($H1$) is at least one population variance σ_1^2 ($1 \leq \sigma \leq k$) is different.

Under H_0, the likelihood function L_0 is

$$L_0 = [1/\ (2\pi\sigma^2)^{N/2}]\ \exp\{-1/2 \textstyle\sum_{ij} ((x_{ij}, - \mu_1)/\sigma_i)^2\},$$

where $N = \sum N_i$

Under H_1, the likelihood function L_1 is

$$L_1 = [1/(2\pi)^{N/2}\ \sigma_1^{N1} \cdots\cdots \sigma_k^{Nk}]\ \exp\{-1/2 \textstyle\sum_{ij} ((x_{ij}, - \mu_1)/\sigma_i)^2\},$$

The joint MLE's of μ_i and σ under H_0 is

$$\mu_i = 1/N_i \textstyle\sum_j x_{ij} = x_i; \quad \sigma^2 = 1/N_i\,(x_{ij}, - x_i)^2 = \sum_i \sigma_i/N$$

In the above the sample variance $vi = S_i/n_i$ with $n_i = N_i - 1$ degrees of freedom.

The joint MLEs of μ_i and σ_i under H_1 is $\mu_i = x_i$ and $\sigma_i^2 = S_i/N_i$

The likelihood ratio L is given by

$$L = (L_1)_{max} / L(1)_{max} = (S_i/N_i)^{Ni/2} \ldots (S_k/N_k)^{Nk/2} / (S/N)^{N/2}$$ where $S = \sum S_i$.

If $L < c$, H_0 may be rejected and we choose c such that the probability $P(L < c / H_0) \leq \alpha$.

For large N, $-2 \log L$ is approximately χ^2 with $(k - 1)$ degrees of freedom. The degrees of freedom $(k - 1)$ are obtained by differencing d.f. under H_0 which is $(k + 1)$ from d.f. under H_1 which is $2k$.

We get $-2\ln L = N\ln S/N - \sum N_i \ln S_i/N_i$.

We can improve the above χ^2 test by replacing N_i by n_i $(=N_i - 1)$ and N by $n(= N - k)$

Thus, $M = c^{-1} [n \ln S/n - \sum n_i \ln S_i/n_i]$ is χ^2 with $(k - 1)$ d.f,

where $c = 1 + [(1/n_i - 1/n)/3(k - 1)]$.

A small value of L will lead to a large role of M since $M = -2 \log L$ and this leads to rejection of H_0. This method works even for small sample size of 3 to 4.

Homogeneity of k-means

We assume that variances $\sigma_1^2 \ldots\ldots \sigma_k^2$ are homogeneous and each equal to σ^2. ANOVA can then easily recognize differences of the treatments (i) having different mean values.

Let x_{ij} be an observation j^{th} member of ith sample, $j = 1\ldots\ldots N_i$ and $i = 1\ldots\ldots k$. A linearly additive mathematical model is

$$x_{ij} = \mu + \alpha_i + \varepsilon_{ij}$$

where for every i, j and ε_{ij} is normally distributed with mean zero and variance σ^2. Summing overall i and j we get

$$\sum x_{ij} = N\mu + \sum N_i \alpha_i + \sum \varepsilon_{ij}$$

where $N = \sum N_i$

We can adjust μ such that $\sum N_i \alpha_i = 0$.

If not, let $\sum N_i \alpha_i = h$, then we can subtract $h/(k N_i)$ for each α_i and add h/N to μ so that vary zero. Let x be the ground mean of x_{ij} $(= N^{-1} = \sum x_{ij})$ and we know that x is an unbiased estimator of population mean. The total sum of squares for all x_{ij} can be split the grand mean and sample mean as

$$x_{ij} - \bar{x} = x_{ij} - \bar{x}_i + \bar{x}_i - \bar{x}$$

Therefore, squaring and summing

$$\sum (x_{ij}-\bar{x})^2 = \sum (x_{ij}-\bar{x})^2 + \sum (x_i-\bar{x})^2 + 2\sum (x_i-\bar{x})(x_{ij}-\bar{x})$$

or, $$St = \sum_j \left(x_{ij}-\bar{x}_i\right)^2 + \sum_i \left(x_i-\bar{x}\right)^2 = S_w + S_a =$$

Since the last term vanishes as $=\sum_j (x_{ij}-\bar{x}_i) = 0$ and $\sum (x_i-\bar{x})$is a constant for the ith sample with same algebra we get

$$S_w = \sum_{ij} x_{ij} - \sum_i((\sum_i x_{ij})^2/N))$$

$$S_a = \sum_i (\sum_j x_{ij})^2 / (N_i - G) \text{ where } G \text{ is } N\bar{x}^2 = (\sum_{ij} x_{ij})^2/N_i.$$

In addition since x_{ij} are normal we have

$$(S_1/\sigma^2) \sim \chi^2_{(N-1)},\ (S_2/\sigma^2) \sim \chi^2_{(N1-1)} \text{ and } (S_a/\sigma^2) \sim \chi^2$$

with $N - 1 - (N - k)$ d.f. and independent of S_w.

Since the expected value of variate is equal to its d.f. , we get

$$E(S_t/\sigma^2) = N-1,\ E(S_w/\sigma^2) = N-k;\ E(S_a/\sigma^2) = k-1,$$

so that $$E(S_t/(N-1)) = \sigma_1^2,\ E(S_w/(N-k)) = \sigma_N^2 \text{ and } E(S_a/(k-1)) = \sigma_a^2.$$

Therefore, ratio of two independent variances $S_a/(k-1)$ and $S_w/(N-k)$ would have F-distribution with $(k-1)$ and $(N-k)$ d.f. The ANOVA Table can be written as follows and F-test can be made to test, differences if any in the treatment means.

Variantions	Sum of square	Degrees of freedom	Mean	F - Test
Among samples	S_a	k-1	Sa/(k-1)	F = (Sa/(k-1)
Within samples	S_w	N-k	Sw/(N-k)	(Sw/(N-k))
Total	$S_t = S_a + S_w$	N-1	St/(N-1)	

When all k samples are of same size r, $N = rk$ and $\sum \alpha_i = 0$ then Su simplified to $S_a = 1/r\ \sum_i\left(\left(\sum_j x_{ij}\right)^2 - G\right)$

Two Way Complete Block Classification

Two factors may be affecting an experiment simultaneously. For example, strength of concrete may depend on (i) cement/sand ratio (ii) time of curing or for an agricultural export production may depend on (i) amount of fertilizers used and (ii) location of blocks along a direction or sandstone porosity may depend on (i) mean size of grains, (ii) sorting coefficients and (iii) grain sphericity.

Each factor is given a few levels and then replicated r times to compute the error variance of the experiments.

The random variable can be called as a linearly additive process and

$$X_{ijk} = \mu + \alpha_i + \beta_j + \gamma_{ij} + \varepsilon_{ijk}$$

where μ is normal mean, γ_{ij} is the interaction between i^{th} and j^{th} levels of two factors.

Again we can adjust $\Sigma\alpha_i = 0$, $\Sigma\beta_{ij} = 0$, $\Sigma\gamma_{ij} = 0$ over each i or j and ε_{ijk} is normal with $E(\varepsilon_{ijk}) = 0$, $V(\varepsilon_{ijk}) = \sigma^2$.

We again obtain,

$$S_t = S_a + S_b + S_{ab} + S_r$$

where $$S_t = \Sigma_{ijk}\, x_{ijk}^2 - G,\ G = N\bar{x}^2 = N^{-1}(\Sigma_{ijk}\, x_{ijk})^2$$

$$S_a = (br)^{-1}\ \Sigma_j\left(\Sigma_{jk}\, x_{ijk}\right)^2 - G$$

$$S_b = (ar)^{-1}\ \Sigma_j\left(\Sigma_{ik}\, x_{jk}\right) - G$$

$$S_{ab} = (r)^{-1}\ \Sigma_{ij}\left(\Sigma_k\, x_{ijk}\right)^2 - S_a - S_b - G$$

$$S_r = (\Sigma_{ijk}\, x_{ijk}) - r^{-1}\ \Sigma_{ij}(\Sigma_k\, x_{ijk})^2$$

The ANOVA Table can be written as F-test mode as follows :

Variation	SS	DF	MS	F
Factor a	S_a	$(a\text{-}1)$	$S_a/(a\text{-}1)$	σ_a^2
Factor b	S_b	$(b\text{-}1)$	$S_b/(b\text{-}1)$	σ_b^2
Inter a × b	S_{ab}	$(a\text{-}1)(b\text{-}1)$	$S_{ab}/(a\text{-}1)(b\text{-}1)$	σ_{ab}^2
Errror	S_r	$ab(r\text{-}1)$	$S_r/ab(r\text{-}1)$	σ^2
Total	S_t	$abr\text{-}1$	$S_t/(abr\text{-}1)$	

The F-tests for a two-way classification is more involved. If the interaction variance is not significant of basis of error variance, then the main effects of factors (a or b) can be easily obtained by ratio σ_a^2/σ^2 (pooled) or σ_b^2/σ^2 (pooled), where σ^2 pooled is the new weighted common error variance obtained by pooling the error and interaction variances as :

$$\sigma^2 \text{ pooled} = (S_r + S_{ab})/[ab\ (r-1) + (a-1)\ (b-1)]$$

However, if interaction variance α_{ab}^2 is significantly greater than error variance σ^2 as obtained by F-ratio = α_{ab}^2/σ^2 with $(a\text{-}1)(b\text{-}1)$ d.f. in numbers $ab(r\text{-}1)$ d.f. in the denominator ah 0.05 level, then σ^2 or α_{ab}^2 cannot be pooled (being non-homogenous variance). Hence the main effects of factor a or b can be obtained by *F*- test by giving α_a^2 or α_b^2 by α_{ab}^2 with (*a*-1) or (*b*-1) d.f. in numerator, respectively and (*a*-1)(*b*-1) if in the denominator. Two-way classifications are easily extended to 3 or 4 way classification.

4.3 ANOVA Models and their Analysis

Theoretically ANOVA models could be of 3 types depending on whether factors are considered fixed (Model I), random (Model II), or partly fixed and partly random (model III). Treatment levels can be considered fixed, for example exact amount of fertilizer used, particular machine operated by man are fixed factors. However it is possible though that the brand name is unknown or residual fertilizer is present, or labour working with any (different) machines in each shift. Thus, the factor is not fixed but is a random variable having finite mean and variance.

Fixed Model (Model I)

For one way classification we have

$x_{ij} = \mu + \alpha_i + \varepsilon_{ij}$, $\quad i = 1, 2, \ldots.. k, \quad j = 1, 2 \ldots.. N_i$ for each *i*.

Mean of ith sample is $\bar{x}_i = \mu + \alpha_i + \bar{\varepsilon}_i$

The normal grand mean is $\bar{x}$, with weights N_i for each $\bar{x}_i$

$$E(\bar{x}) = \mu, \text{ so } \bar{x} \text{ is, } (\bar{x} = \mu + \bar{\varepsilon}) \text{ an unbiased estimate of } \mu.$$

The expectation of x_i is $\mu + \alpha_i$, its variance is σ^2/N_i $\bar{x}_i - \bar{x}$ is and unbiased estimate

$$S_b = \Sigma_i N_i(\bar{x}_i - \bar{x})^2 \text{ of } \alpha_i$$

with
$$E(S_b) = (k-1)\sigma^2 + \Sigma_i N_i\alpha_i^2$$

Thus, $S_b/(k-1)$ is greater than σ^2 in turn all a_i are zero so we should use one tailed *F*-test.

$$E(S_t) = (N\text{-}1)\sigma^2 + \Sigma_i N_i\alpha_i^2$$

$$E(S_w) = (N-k)\sigma^2$$

which shows the fixed treatment effects do not enter in to the mean sure within samples.

For a two-way classification with fixed effects, we have

$$E(S_a) = (a-1)\sigma^2 + br\,\Sigma_i \alpha_i^2$$

$$E(S_b) = (b-1)\sigma^2 + ar\,\Sigma_j \beta_j^2$$

$$E(S_{ab}) = (a-1)(b-1)\sigma^2 + r\,\Sigma_{ij}\gamma_{ij}^2$$

$$E(S_r) = ab\,(r-1)\sigma^2$$
$$E(S_t) = (abr-1)\sigma^2 + br\,\Sigma_i \alpha_i^2 + ar\,\Sigma_j \beta_j^2 + r\,\Sigma_{ij} \gamma_\mu^2$$

The unbiased estimators of α_i, β_j, γ_{ij} and μ are

$$\hat{\alpha}_i = \bar{x}_i - \bar{x}, \quad i = 1, 2 \ldots\ldots a$$

$$\hat{\beta}_j = \bar{x}_j - \bar{x}, \; j = 1, 2 \ldots\ldots b$$

$$\hat{\gamma}_{ij} = \bar{x}_{ij.} - \bar{x}_{i..} - \bar{x}_{j.} + \bar{x}$$

$$\hat{\mu} = \bar{x}$$

Variable Factor (Model II)

Here the effects indicating interactions are treated as random variates that are normal and independent distributions. Hence α_i is random with expectation zero and variance σ_a^2 in a one-way classification.

$$x_{ij} = \mu_i + \varepsilon_{ij} = \mu + \alpha_i + \varepsilon_{ij}$$
$$V(x_{ij}) = \sigma_a^2 + \sigma^2$$

where σ_a^2 and σ^2 are called *components of variance.*

If two observations in the *same* sample are available then

$$\text{cov}\,(x_{ij}, x_{ij}) = E\,(\alpha_i^2)$$
$$= \sigma_a^2$$

The intra-class correlation coefficient is fixed by $\sigma_a^2/(\sigma_a^2 + \sigma^2)$.

The null hypothesis is $\sigma_a^2 = 0$, which imply $m_u = \mu$ for all u.

Assuming all samples are replicated r times the SS between samples and SS within samples are

$$S_b = r\,\Sigma_i (\bar{x}_i - \bar{x})^2$$
$$= r\Sigma_i\,[(\mu + \mu_i + \hat{\varepsilon}_{i.}) - (\mu + \bar{\alpha} + \bar{\varepsilon})]^2$$
$$S_w = \Sigma_{ij} (\bar{x}_{ij} - \bar{x}_i)^2$$
$$= \Sigma_{ij}\,(\varepsilon_{ij} - \bar{\varepsilon}_i)^2$$

If the ε_{ij} is normal with expectation zero and variance σ^2, S_w/σ^2 has a distribution with $k(r-1)$ d.f., so that

$$E\,[S_w / k(r-1)] = \sigma^2.$$

If also the α_i are normal with variance α_a^2, and if we write η_i for variable $\alpha_i + \bar{\varepsilon}_{i.}$, then

$$S_b = r\Sigma_i(\eta_i - \bar{\eta})^2$$

and the η_i are normal with mean zero and variance $\sigma_a^2 + (\sigma^2/r)$.

Therefore, $S_b/[(\sigma_a^2 + (\sigma^2/r)\ r]$ is a chi-square variate with $(k-1)$ d.f, so that

$$E\ [S_b / (k-1)] = r\ \sigma_a^2 + \sigma^2.$$

The Sums of Squares (SS) for among groups is S_b and within group is S_w are statistically independent. So, null hypothesis can be tested by ordinary F-test as σ_a^2 / σ_w^2. The power of F-test is σ_a^2 / σ^2.

For a two-way classification, $x_{ijk} = \mu + \alpha_i + \beta_j + \gamma_{ij} + \gamma_{ijk}$ where μ, $\alpha_i + \beta_j$, γ_{ij} and ε_{ijk} are widely and normally distributed with mean zero and variances σ_a^2, σ_β^2, σ_γ^2 and σ^2 respectively,

The value of $\quad V(x_{ijk}) = \sigma_\alpha^2 + \sigma_\beta^2 + \sigma_\gamma^2 + \sigma^2$

$$S_a = br\ \Sigma_i\ (x_{i..} - x)^2$$

$$= br\ \Sigma_i = [(\alpha_i - \bar{\alpha}_i + \bar{\gamma}_{i,} - \bar{\gamma} + \bar{\varepsilon}_{i.} - \bar{\varepsilon})]^2$$

If $\eta_i = \alpha_i + \bar{\gamma}_i + \bar{\varepsilon}_{i..}$, then η_i is normal with mean zero and variance

$$\sigma_a^2 + \sigma_\gamma^2/b + \sigma^2/(br).$$

Its mean over a values of i is

$$\bar{\eta} = \bar{\alpha} + \bar{\gamma} + \bar{\varepsilon}$$

Therefore, $\quad [\Sigma_i(\eta_i - \bar{\eta})^2]/[\sigma_a^2 + \sigma_\gamma^2/b + \sigma^2/(br)]$ is χ^2 with $(a-1)$ d.f.

and $\quad E\ [S_a/(a-1)] = br\ [\sigma_a^2 + \sigma_\gamma^2/b + \sigma^2/(br)]$

$$= br\ \sigma_a^2 + r\sigma_\gamma^2 + \sigma^2$$

In the same way $E[S_b/(b-1)] = ar\sigma_\beta^2 + r\sigma_\gamma^2 + \sigma^2$

$$E[S_{ab}/(a-1)(b-1)] = r\sigma_\gamma^2 + \sigma^2$$

$$E[S_r/ab(r-1)] = \sigma^2$$

The four SS are statistically independent and so ordinary F-test for hypothesis $\sigma_a^2 = 0$, $\sigma_\beta^2 = 0$, $\sigma_\gamma^2 = 0$ may be carried out. Here interaction components of variance are found in main effects.

Mixed Models (Model III)

We can illustrate this with one factor and fixed as the other factor random. Suppose we are interested in a factory production with the performance of individual machines of different makes (fixed effects) where as the labourers are random sample for a large population of labour force.

Let x_{ijk} be the output of jth worker on k^{th} day that has assigned to the ith machine ($i = 1 \ldots a$, $j = 1, 2 \ldots b$, $k = 1, 2 \ldots.. r$). The days are regarded as replicate outputs and are interested in fixed effects α_i and random effects β_j then,

$$x_{ijk} = m_{ij} + \varepsilon_{ijk}.$$

where m_{ij} is "*true*" mean effect of jth worker in the ith make and ε_{ijk} are normal with mean zero and variance σ^2.

We can regard m_{ij} as a further value of the random variable M_i which reflect the mean output of a worker selected at random on machine numbered i.

$E(M_i) = \mu$ and the arithmetic mean of μ_i over i machine be denoted by μ then,

$$\mu_i = E(M_i), \qquad \mu = \mu = \Sigma\mu_i/a$$

The main effect of ith machine is $\alpha_i = \mu_i - \mu$, $\Sigma\alpha_i = 0$

Suppose m_{iw} is the value of M_i for any worker is the labour force. The true mean for this worker is average of m_{iw} over i machines and the main effect β_w of worker w is

$$\beta_w = \overline{m}_{\cdot w} - E(\overline{M}.)$$

$$= \overline{m}_{\cdot w} - \mu$$

This random variable has mean zero and variance σ^2_β. The main effect of worker w, specific to i^{th} marker, is $m_{iw} - E(M_i) = m_{iw} - \mu_i$

This is interaction of ith machine with particular worker w

$$\gamma_{iw} = \mu_{iw} - \mu_i - \overline{m}_w + \mu$$

For each i this is a random variable with median zero as variance σ^2_γ.

Also $\qquad \Sigma\gamma_{iw} = 0$ for each w.

Therefore, we get $m_{iw} = \mu + \alpha_i + \beta_w + \gamma_{iw}$

And for any worker w and any day d

$$x_{iwd} = \mu + \alpha_i + \beta_w + \gamma_{iw} + \varepsilon_{iwd}$$

where ε_{iwd} are independent of each other and of β_w and γ_{iw} and have a common variance σ^2 . The γ_{iw}, for different values of i are not necessarily independent of each other or of β_w, but have covariances depending on those of the random variables M_i.

The b workers used may be regarded as a random sample of labour and γ days form a random sample of all possible days.

The x_{ijk} has the same form as x_{iwd}, but j now takes values 1, 2, ... b, and k takes values 1, 2, ... r.

$$x_{ijk} = \mu + \alpha_i + \beta_j + \gamma_{ij} + \varepsilon_{ijk}$$

We define $(a-1)\ \sigma_a^2 = \sum_i \alpha_i^2$ but since α_i are fixed effects σ_a^2 is not the variance of a random variable. Also we have $(a-1)\ \sigma_\gamma^2 = \sum_i \alpha_{\gamma_i}^2$.

The division of total sums of squares may be made as

$$S_t = S_a + S_b + S_{ab} + S_r$$

where $S_a = br \sum_i (\bar{x}_{i.} - \bar{x})^2$

$$S_b = ar \sum_j (\bar{x}_{.j.} - \bar{x})^2$$

$$S_{ab} = r \sum_{ij} (\bar{x}_{ij.} - \bar{x}_{.j.} - \bar{x})^2$$

$$S_r = \sum_{ijk} (x_{ijk} - \bar{x}_{ij.})^2 \sum_{ijk} (x - \bar{x}_{ij})^2$$

We get, $E\,[S_a/(a-1)] = br\,\sigma_a^2 + r\,\sigma_a^2 + \sigma^2$

$$S_b = ar \sum_j (b_j - \bar{\beta}) + (\varepsilon_{.j.} - \bar{\varepsilon})^2$$

$$= ar \sum_j (\beta_j + \bar{\beta}^2) + ar \sum_j (\bar{\varepsilon}_{..j.} - \varepsilon)^2 + 2ar \sum_j (\beta_j - \beta)\ (\varepsilon_{.j} - \varepsilon)$$

$$= (b-1)\ \sigma_\beta^2 + (b-1)\ \sigma^2/ar + 0 \text{ (as } \beta_j, \textstyle\sum_{ijk} \text{ are independent)}$$

Therefore, $E\,(S_b) = ar\,(b-1)\ \sigma_\beta^2 + (b-1)\ \sigma^2$

$$E\,[S_b/(b-1)] = ar\,\sigma_\beta^2 + \sigma^2$$

Similarly, $S_{ab} = r \sum_{ij} (\gamma_{ij} - \gamma_{i.} + \bar{\varepsilon}_{ij.} - \bar{\varepsilon}_i - \bar{\varepsilon}_{i.} - \varepsilon_{.j.} + \varepsilon)$

$$= r\,(a-1)\ (b-1)\ \sigma_\gamma^2 + (a-1)(b-1)\ \sigma^2.$$

So $E\,[S_{ab}/(a-1)\ (b-1)] = r\,\sigma_\gamma^2 + \sigma^2$.

and finally $S_r = ab \sum_k (\varepsilon_{ijk} - \bar{\varepsilon}_{ij.})^2$ and $E\,[S_r] = ab(r-1)\ \sigma^2$

So $E\,[S_r/ab\,(r-1)] = \sigma^2$

The ANOVA table for model III (mixed) is given by:

A effects (fixed)	$a - 1$	$S_a/(a-1) = \sigma_a^2$	$br\,\sigma^2 + r\,\sigma_\gamma^2 + \sigma^2$
B – effects (random)	$b - 1$	$S_a/(b-1) = \sigma_b^2$	$a'\,\sigma_\beta^2 + \sigma^2$
A × B effects (random)	$(a-1)(b-1)$	$S_{ab}/(a-1)(u-1) = \sigma_{ab}^2$	$r\,\sigma_\gamma^2 + \sigma^2$
Error	$ab\,(r-1)$	$S_r/ab(r-1) = \sigma^2$	σ^2
Total	$ab\,r - 1$		

The four sum of square are pairwise independent, see expressions for S_b and S_{ab}. Therefore, we can test for interaction by *F*-tests ratio σ_{ab}^2/σ_b^2, B-effects by σ_b^2/σ^2, and A-effects by σ_b^2/σ_{ab}^2. The estimates of μ is the over-all mean $\bar{x}$, and estimators of α_i, β_j and γ_{ij} are respectively

$$\hat{\alpha}_i = \bar{x}_{i..} - \bar{x}\,;\ \hat{\beta}_j = \bar{x}_{.j.} - \bar{x}\,;\ \hat{\gamma}_{ij} = \bar{x}_{ij} - \bar{x}_{i.}\,\bar{x}_{.j.} + \bar{x}$$

In addition to complete block designs as given above, we can have incomplete block designs, nested (or hierarchical) models, Latin Square Designs etc.

Violations of ANOVA Assumptions

The assumptions for ANOVA analysis are (a) additivity implying absence of interaction terms (b) independence of error terms (c) normality of distribution of error terms, (d) constancy of variance of error terms, whatever may be the amplitude of the main effects. Together these provide severe restrictions for ANOVA analysis and inference.

Non-normality does not introduce bias into point estimates of parameters or components of variance. It affects the validity of *F*-tests, as *MS* need not be independent chi-square distributions. If x_i and $x_i + 1$ are not independent but serially are correlated (ρ), then

$$E(\bar{x}) = \mu;\ V(\bar{x}) = \sigma^2/N(1 + (1 - 1/N));$$
$$E(2\rho) = \sigma^2/(1 - (2\rho/N))$$

So there is a good chance of serious error if ρ is neglected. Usually nonlinear transformation of random variable (say logarithmic) is used to induce homogeneity of variances and this may simultaneously bring about normality to the error variables. However, transformation must not destroy the additivity of effects. If *F*-test shows significance for main effects, we would like to know which level (levels) have different mean value(s). We can use Newman–Keuls procedure.

Let a be the number of levels of the features and let m_e with ν d.f. be the MS used in denominator of *F*-test. Let *r* be the replication. The observed means are arranged from smallest to the greatest and the test is carried out in sub groups of *p* successive means beginning with the whole set ($p = a$).

The studentized range of p observation with actual range R and having population variance σ^2 is R/S_v, where S^2_v is an independent estimate of σ^2 based on v d.f. Thus $P\,(R/s_v \leq \theta)$ can be compared with percentage points of studentized range(Biometrika Tables).

4.4 Latin Square Design

In a complete experimental design for three factors with m (=3) levels each and one replication per cell we have to obtain $3^3 = 27$ data. Latin square is an incomplete design with same number of levels (=3), but observations are made on $3^2 = 9$ experiments, with obvious economy. However, these designs do not allow for interactions. In Latin squares, the rows and columns are permuted so that each level comes only once in row/column. For example, a 3 × 3 design in a natural order is given by

	1	2	3
1	A	B	C
2	B	C	A
3	C	A	B

which can be permuted to give other designs.

As the number of levels is increased the number of possible designs becomes very large to have much flexibility.The mathematical model is given by

$$x_{ijk} = \mu + \alpha_i + \beta_j + \gamma_k + \varepsilon_{ijk}$$

where i,j,k take values from 1 to m (=number of levels) but only m^2 sets of triples (i,j,k) are permissible. Here α_i are the main effects of first factor (say treatment), β_j for the rows and γ_k for the columns. The ε_{ijk} are independently normal with mean zero and variance σ^2. The mean values of α_i, β_j, γ_k, are adjusted to zero as in ANOVA designs. The null hypotheses to be tested are :

(a) all α_i, = 0

(b) all β_j, = 0

(c) all γ_k = 0

Assuming Model-I (fixed effects) we have

$$S_t = S_a + S_b + S_c + S_e$$

with S_a, S_b, S_c, S_e computed as usual.

For example, $S_t = \sum_{ijk}(x_{ijk} - \bar{x})^2 = \sum x^2_{ijk} - m^2\,\bar{x}^2 (-G)$

and $S_a = m\sum_i \bar{x}^2_{i.} - G$ etc.

ANOVA Table for Latin Square is as follows :

Variation	S.S	DF	MS	E(M.S)
Treatment (A)	S_a	$m-1$	$S_a/m-1$	$\sigma^2 + m\,\sigma_a^2$
Rows (B)	S_b	$m-1$	$S_b/m-1$	$\sigma^2 + m\,\sigma_b^2$
Columns (C)	S_c	$m-1$	$S_c/m-1$	$\sigma^2 + m\,\sigma_c^2$
Error	S_e	(m^2-3m+2)	$S_e/(m^2-3m+2)$	σ^2
Total	S_t	m^2-1		

Since the first four SS are independent, we can make *F*-test for the above three null hypotheses. The decisions assume no interactions among A, B, C. However, if there are interactions between these factors, then interpretation of results would be difficult.

We may consider biogeochemical leaching of in situ gold for gold ores using three types of bacteria. The numbers of in situ squares are 9 arranged along three rows and three columns but only one type of bacteria is put in each square so that no row and column has more than one strain. The result of this Latin square design will give us clues as to which strain performs better or yields greater leaching as well as if there are any differences in gold leaching along rows or along columns. These results would help in large-scale economical designing of the gold leaching operation in the mine.

4.5 Regression and Correlation for Pairs of Variables

We wish to measure the relationship between two random variables (*X* and *Y*) both of which can be measured in each member of a given population. For simplicity, *X* and *Y* are assumed to be continuous random variables but these analyses can be extended to discrete random variables by replacing integrals by sums.

Let the joint probability density function be $f(x,y)$ and marginal density

$$g(x) = \int_{-\infty}^{\infty} f(x,y)dy \text{ and } h(y) = \int_{-\infty}^{\infty} f(x,y)\,dx.$$

The conditional-density of *Y* for a given value x of *X* is

$$f(y \mid x) = f(x, y)/g(x)$$

Similarly, $$f(x \mid y) = f(x, y)/h(y).$$

The expectation of *Y*, for given *X*, is defined as

$$\eta_x = E[Y \mid X] = \int_{-\infty}^{\infty} y\,f(y \mid x)\,dy$$ and it is a function of x.

The graph of η_x as a function of x is called the true regression curve of *Y* on *X*. If this regression is linear, the group is straight line with the equation $\eta_x = \alpha + \beta x$. The parameter β is the true regression coefficient of *Y* on *X* and is the slope of the straight line. The other parameter a is the intercept on the axis of Y at $x = 0$. The variance of Y conditional on x is

$$\sigma^2_{Y|x} = \gamma\,(Y \mid X) = \int (y - \eta_x)^2 f(y|\,x)dy.$$

This is generally a function of x, although sometimes it could be independent of x.

The regression parameters can be represented by moments of the joint distribution of X and Y. Let the two means be

$$\mu_x = \int_{-\infty}^{\infty} y(x)x\,dx = \iint x\,f(x, y)dy\,dx$$

$$\mu_Y = \int_{-\infty}^{\infty} h(y)\,ydy = \iint y\,f(x, y)dx\,dy$$

and the two variances and covariance are

$$\sigma^2_x = \int_{-\infty}^{\infty} y(x)(x - \mu_x)^2\,dx$$

$$\sigma^2_Y = \int_{-\infty}^{\infty} h(y)\,(y - \mu_y)^2\,dy$$

$$\sigma_{xy} = \int_{-\infty}^{\infty} f(x, y)(x - \mu_x)(y - \mu_y)dy\,dx$$

We can easily obtain $g(x)\ \eta_x = \int_{-\infty}^{\infty} y\,f(x, y)dy$ so that on writing $\eta_x = (\alpha + \beta x)$ and integrating,

$$\int_{-\infty}^{\infty} (\alpha + \beta x)\,g(x)dx = \int_{-\infty}^{\infty}\int_{-\infty}^{\infty} y\,f(x, y)dy\,dx \text{ or } \alpha + \beta\mu_x = \mu_y.$$

Multiplying $g(x)$ by x we obtain

$$\int_{-\infty}^{\infty} (\alpha + \beta x^2)\,g(x)dx = \int_{-\infty}^{\infty}\int_{-\infty}^{\infty} y\,f(x, y)dy\,dx$$

which becomes $\alpha + \beta\mu'_{2x} = \pi_{xy}$

where μ'_{2x} is the second moment of X about the origin and the π_{xy} is the product moment of X and Y about the origin.

Multiplying by μ_x we get

$$\beta\,(\mu'_{2x} - \mu^2_x) = \pi_{xy} - \mu_x\mu_y \text{ or } \beta\,.\,\sigma^2_x = \sigma^2_{xy}$$

Therefore, we have

$$\beta = \sigma_{xy}/\sigma^2_x = \rho_{xy}\,\sigma_y/\sigma_x$$

where ρ_{xy} (Pearson's correlation coefficient) $= \sigma_{xy}/(\sigma_x\,\sigma_y)$.

From equation above, $\alpha = \mu_y - \beta\mu_x$, so that both β and α are expressible in terms of variances and covariances of X and Y.

Equation to regression line is

$$\eta_x - \mu_x = \beta\,(x - \mu_x) = \rho_{xy}\,\sigma_y\,((x - \mu_x)/\sigma_x)$$

which shows that the regression line passes through $(\mu_x\,.\,\mu_y)$.

Interchanging the variable a similar regression line for a given value of y is

$$\zeta_y - \mu_x = \rho_{xy}\, \sigma_x((y-\mu_y)/\sigma_y).$$

This line also passes through the point (μ_x, μ_y) but does not coincide with the previous regression line.

The variance of estimate of Y for given x is

$$\sigma_{Ye}^2 = \int_{-\infty}^{\infty} \sigma_{y/x}^2\, g(x)\, dx.$$

and this becomes average variability of Y and the regression line of Y on X.

We compute $\quad (y-\eta_x)^2 = \{y-\mu_y-\beta(x-\mu_x)\}^2$

$$= (y-\mu_y)^2 + \beta^2(x-\mu_x)^2 - 2\beta\,(y-\mu_y)(x-\mu_x)$$

and we get

$$\sigma_{Ye}^2 = \int_{-\infty}^{\infty}\int_{-\infty}^{\infty} [(y-\mu_y)^2 + \beta^2(x-\mu_x)^2\, 2\beta(y-\mu_y)(x-\mu_x)]f(x,y)dxdy$$

$$= \sigma_x^2 + \beta^2\,\sigma_y^2 + 2\,\beta\,\sigma_{xy}.$$

Since $\quad \beta^2\sigma_x^2 = \sigma_{xy}^2/\sigma_x^2 = \beta\sigma_{xy} = \rho_{xy}^2\sigma_y^2$, we obtain

$$\sigma_{ye}^2 = \sigma_y^2\left(1-\rho_{xy}^2\right)$$

Since both σ_{ye}^2 and σ_y^2 cannot be negative, we have $\rho_{xy}^2 \le 1$.

Hence $-1 \le \rho_{xy} \le 1$ for all values of X and Y.

If X and Y are independent $\rho_{xy}^2 = 0$ (but the converse is not necessarily true).

If Y is proportional to X then $\sigma_{ye}^2 = 0$ and $\rho_{xy} = \pm 1$ (perfectly correlated), which are extreme cases.

Bivariate Normal Density

An important special case of a two-variate distribution is that the joint probability density is bivariate normal. i.e.

$$f(x,y) = \left[\frac{1}{\left(2\pi\pi_x\sigma_y\left(1-\rho^2\right)^{1/2}\right)}\right] e^{-1/\left(2-\left(1-\rho^2\right)\right)} \left[\left((x-\mu_x)/\sigma_x\right)^2 + \left((y-\mu_y)/\sigma_y\right)^2\right]$$

$$- 2\,\rho\,[((x-\mu_x)\,/\,\sigma_x)\,((y-\mu_y)\,/\,\sigma_y)]$$

where $\quad \rho = \rho_{xy}.$

In terms of standardized variable $z = (x-\mu_x)/\sigma_x$ and $v = (y-\mu_y)/\sigma_y$, we get bivariate normal surface,

$$g(z, v) = (2\pi)^{-1}(1 - \rho^2)^{-1/2} \exp [-(z^2 + v^2 - 2\rho z v)/2 (1 - \rho^2)]$$

The marginal densities are standard normal distribution as follows

$$f(z) = \int_{-\infty}^{\infty} g(z, v)\, dv = (2\pi)^{-1/2}\, e^{-z^{2/2}}$$

$$h(v) = \int_{-\infty}^{\infty} g(z, v)\, dz = (2\pi)^{-1/2}\, e^{-z^{2/2}}$$

The conditional probability for v for a given z is

$$g(v|z) = g(z, v)/f(z)$$
$$= [(2\pi)(1 - \rho^2)^{-1/2} \exp [-(v - \rho z)^2/2(1 - \rho^2)]$$

The overall expectation of v for a given z is $\eta_z = \int_{-\infty}^{\infty} v g(v|z)\, dv = \rho z$

This is straight line through origin with slope ρ. The regression is linear and this holds also for the regression $\xi_v = \rho v$. Each array of v's for a given z, has variance

$$\sigma_{v|z} = \int_{-\infty}^{\infty} (v - \rho z)^2\, g(v|z)\, dv$$

and integrating we obtain,

$$\sigma^2_{v|z} = 1 - \rho^2$$

Transforming back to integral variables (x, y) we have

$$\sigma^2_{y|x} = \sigma^2_y (1 - \rho^2)$$

and the variance is therefore independent of x. The weighted average σ^2_{ye} is then same as $\sigma^2_{y/x}$ for all x. A similar property holds for y-arrays of X for given values of y.

In the z-v plane the two regression lines η_z and ξ_v have the same slope ρ. A section through the bivariate normal surface by a horizontal plane g = constant, is an ellipse with equation $z^2 + v^2 - 2\rho z v = c$. The tangents at the points of intersection of regression lines with the ellipse are parallel to the co-ordinate axes X, Y.

Linear Regression from a Sample

In practice X, Y may be random, or only one of them or neither. One of them is time or spatial coordinate giving rise to temperature rises, stock market prices, assay values in a drill core in which sample values are not necessarily independent (but correlated) (Fig. 4.1 & 4.2).

Also often values of X are pre-selected rather than random and one or both X, Y may be subject to experimental errors. The time relation between X and Y is a functional one and unless Y, X are random, there is little sense in speaking of their distribution, expectation or variance etc.

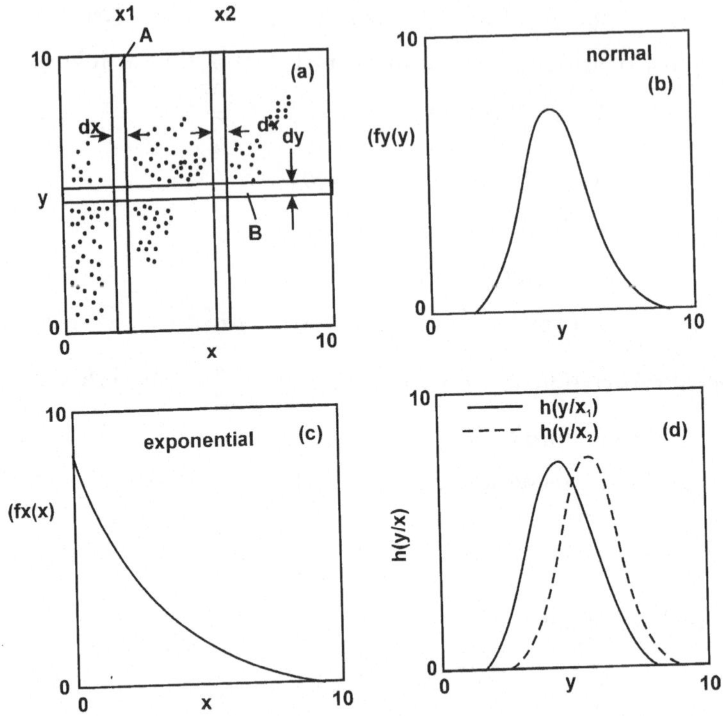

Fig. 4.1 Joint distribution of sample points for $f(x, y)$ when marginals $f_Y(y)$ is Normal and $f_X(x)$ is Exponential. The conditional p.d.f $h(y|x_1)$ and $h(y|x_2)$ are different. Hence, y is not independent of x

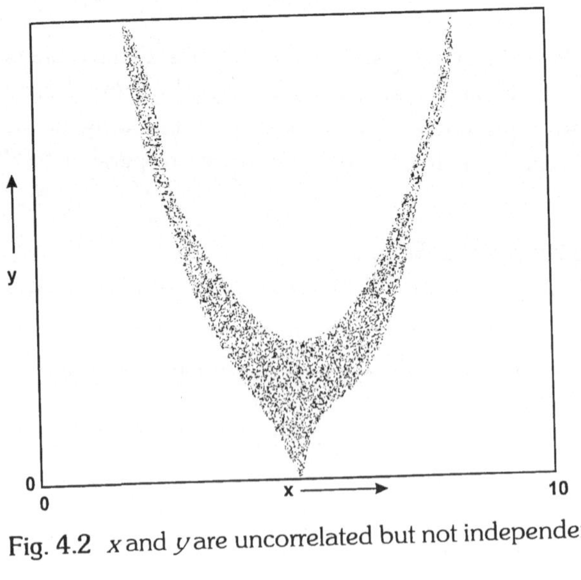

Fig. 4.2 x and y are uncorrelated but not independent

In the normal regression problem, Y is a random variable and X may either be random or fixed. The classical situation is that when both X and Y are random a sample is selected randomly for the population and X, Y values measured in each of N selected items. We assume that for a given value x of X, Y has the linear regression form

$$Y = \alpha + \beta x + \varepsilon$$

where random error ε is normally distributed with mean zero and variance σ^2.

The relation $Y = \alpha + \beta x$ which really is $E(Y|X = x) = \alpha + \beta x$ is called a structural relation. It is this underlying relationship which is disturbed, in an actual sample, by the sampling fluctuation of Y and by errors of measurement of both X and Y.

Assuming X and Y are measured precisely (without much errors), so that ε now represents sampling fluctuations. If a straight line is fitted to the N sample values of X and Y, we can get parameters of α and β of the regression line. Shortly it will be demonstrated that least square method of fitting yields unbiased estimates of α and β.

Let the sample values be (x_i, y_i), $i = 1, 2 \ldots N$. We have to find out the best fit line i.e. best estimate of Y for a given x of X and to estimate standard error of parameters. Let least square fit line be $y_c = \alpha + kx$, such that $S = \sum_i e_i^2 = \sum_i (y_i - a - bx_i)^2$ is minimum. The true regression line is fitted by $\eta = \alpha + \beta x$ with error ε_i.

Differentiating and equating to zero, we get $\partial s / \partial a = 0; \quad \partial s / \partial b = 0$.

We solve simultaneously for the two equations for the two unknown parameters to obtain,

$$\Sigma_i (y_i - a - bx_i) = 0,$$
$$\Sigma_i x_i (y - a - bx_i) = 0$$

or

$$Na + \Sigma_i x_i b = \Sigma_i y_i$$
$$\Sigma_i x_i a + \Sigma_i x_i^2 b = \Sigma_i x_i y_i$$
$$b = s_{xy}/s_x^2, \ a = \bar{y} - b\bar{x}$$

where $\bar{x} = \Sigma_i x_i/N, \quad \bar{y} = \Sigma_i y_i/N, \quad (N\text{-}1)\, s_x^2 = \Sigma_i x_i^2 - N\bar{x}^2$ and $(N\text{-}1)\, s_{xy} = \Sigma_i x_i y_i - N\bar{x}\bar{y}$.

Here s_x^2 is sample variance of X and s_{xy} is the sample covariance of X and Y. The Pearson effect of covariance for the sample is

$$r_{xy} = s_{xy}/s_x s_y = bs_x/s_y$$

so that the least square linearity may written as

$$y_c - \bar{y} = b(x-\bar{x}) = (rs_y/s_x)(x-\bar{x}) \quad \text{where } r = r_{xy}.$$

Integrating X and Y we get best fitted least squares line

$$x_C - \bar{x} = b'(y-\bar{y}) = (rs_x/s_y)(y-\bar{y})$$

where $$b' = s_{xy}/s_y^2.$$

We note that $bb' = r^2$, so that r is the geometric mean of two slopes (one for x-axis, one for y-axis). These two regression lines intersect at $(\bar{x}, \bar{y})$. In most practical situation one of the variables, which we want to predict, is denoted as Y to compute the first regression line and the second regression line $E(X|Y=y)$ is rarely needed.

Calculation of b and r requires two variances and the covariances of X and Y. It is easy to obtain in computational formula as

$$b = [N\,\Sigma xy - \Sigma x \Sigma y] / [N\,\Sigma x^2 - (\Sigma x)^2]$$

$$r^2 = [N\,\Sigma xy - \Sigma x.\Sigma y]^2 / [(N\,\Sigma x^2 - (\Sigma x)^2)(N\Sigma y^2 - (\Sigma y)^2)]$$

Since the correlation coefficient is a dimensionless slope, it is independent of the units of measurements of X and Y. Therefore, correlation lines are invariant under change of scale of X and Y.

Variance about the Regression Line

When a, b are chosen by least square, the sum of the error is minimized or

$$S_{min} = \Sigma_i\, e_i^2$$

$$= \Sigma_i [y_i - \bar{y} - b(x-\bar{x})]^2$$

$$= (N\text{-}1)\,(s_y^2 + b^2 s_x^2 - 2bs_{xy})$$

$$= (N\text{-}1)\, s_y^2 (1 - r^2)$$

This suggests the $S_{min}/(N-1)$ is an estimator of σ_{ye}^2. However a better (unbiased) estimator is $S_{min}/(N-2)$.

If β is true regression slope and e_b is error or sample function of b, then

$$b = \beta + \varepsilon_b$$

$$e_b = \Sigma_i \frac{\varepsilon_i(x-\bar{x})}{(N-1)s_x^2}$$

Since $E(\varepsilon_i) = 0$ for all i, it follows then $E(e_b) = 0$ and therefore, b is an unbiased estimator of β. Also ε_i are homoscedastic with variance σ^2.

Therefore, $$V(b) = V(e_b) = \sigma^2 \Sigma_i \frac{(x_i-\bar{x})^2}{(N-1)s_x^4}$$

$$= \sigma^2/(N\text{-}1)\ (s_x^2)$$

Moreover, if ε_i is normal for each i, so is b,with expectation β and variance $\sigma^2/(N\text{-}1)\ (s_x^2)$.

We also have $\varepsilon_i - e_i = (a-\alpha) + (b-\beta)x_i$

or $(a-\alpha) + (b-\beta)\,\bar{x} = \bar{\varepsilon}$

$\varepsilon_i - e_i = \bar{\varepsilon} + e_b\,(x_i - \bar{x})$

For a fixed x value, $y_c - \eta = (a-\alpha) + (b-\beta)x$

$= \bar{\varepsilon} + e_b\,(x_i - \bar{x})$

$$= \Sigma_i\ \varepsilon_i \left[\frac{1}{N} + \frac{(x-\bar{x})(x_i-\bar{x})}{(N-1)\,s_x^2}\right]^2$$

$$= \sigma^2 \left[\frac{1}{N} + \frac{(x-\bar{x})(x_i-\bar{x})}{(N-1)\,s_x^2}\right]$$

Since ε_i are independent with mean zero and common variance σ^2 we get,

$$E(y_c) = \eta$$

and $$V(y_c) = \sigma^2\ \Sigma_i \left|\frac{1}{N} + \frac{(x-\bar{x})(x_i-\bar{x})}{(N-1)\,s_x^2}\right|^2$$

$$= \sigma^2 \left[\frac{1}{N} + \frac{(x-\bar{x})(x_i-\bar{x})}{(N-1)s_x^2}\right]^2 \quad \text{as } \Sigma_i(x_i-\bar{x})^2 = (N-1)\ s_x^2$$

S_{min}/σ^2 is distributed as χ^2 with (N-2) d.f

$$E\,(S_{min}/\sigma^2) = N - 2$$

or $$E\,(S_{min}/(N\text{-}2) = \sigma^2$$

An unbiased estimate of σ^2 is therefore,

$$\hat{\sigma}^2 = (S_{min}/(N\text{-}2))$$

$$= \frac{N-1}{N-2}\left[s_y^2\left\{\left(1-r^2\right)\right\}\right]$$

So, $$\hat{v}\left(y_c\right) = \hat{\sigma}^2\left[\frac{1}{N} + \frac{\left(x-\bar{x}\right)^2}{\left(N-1\right)s_x^2}\right]$$

which is minimum at $x = \bar{x}$, and increase for higher or lower values from $\bar{x}$.

Variance of Y about estimated regression line for a new value x of X is sometimes of interest.

$$V(Y - y_c / x) = V(Y / x) + V(y_c)$$

$$= \sigma^2\left[1 + \frac{1}{N} + \frac{\left(x-\bar{x}\right)^2}{\left(N-1\right)s_x^2}\right]$$

We substitute in $\hat{\sigma}^2$ place of σ^2 to obtain the unbiased estimated variance and its square root is called standard error of estimate.

4.6 Confidence Limits of Parameters and Predicted *Y* Value

Since b is a normal with mean β and variance $\sigma^2/[(N - 1)(s_x^2)]$ and an independent unbiased estimator $\sigma^2$with N-2 d.f.is obtained $(b - \beta)\ [(N - 1)(s_x^2) / \hat{\sigma}^2]^{1/2}$ has Student-t distribution with $(N - 2)$ d.f.. So we can get the confidence interval for β as,

$$\beta = b \pm t_\alpha \left(\frac{s_y^2\left(1-r^2\right)}{\left(N-2\right)s_x^2}\right)^{1/2}$$

$$= b \pm t_\alpha \frac{b}{r}\left(\frac{\left(1-r^2\right)}{\left(N-2\right)}\right)^{1/2}$$

The variance of a is $\sigma^2 \left[\frac{1}{N} + \frac{\bar{x}^2}{(N-1)\, s_x^2} \right]$ which gives the confidence intervals. In most applications the slope (β) is important rather than intercept (α). It should be noted that the confidence limits are narrowest at the center of scatter points $(\bar{x}, \bar{y})$ and increases continuously away from the center point in both (higher and lower) directions.

If X values are pre-selected then it is not a random variable and sometimes X values are not under the control of the experimenter. The problem really is that Y variable is a function of a mathematical variable x. The linear model is then $Y = \alpha + \beta x + \varepsilon_x$, where ε_x are normal with mean zero and a common variance σ^2 for all x and ε_x are independent for different x values. The MLE for α and β are again a and b the least square linear regression parameters respectively. However, s_x^2 is not an estimator of variance of X but represents $\sum (x_i - \bar{x})^2 / (N-1)$, where $\bar{x} = \sum x_i / N$. But, in some practical applications it may become necessary to predict X for a given value of Y although X is not random. Here we insert the regression equation and obtain $x = (y_c - a)/b$. The 100 (1–α)% confidence limit can then be obtained by allowing $y = y_c$ (horizontal) line intersect the confidence belt curves at x_1 and x_2. This is of practical importance when numerical value of b is large (not close to zero).

MLE of Regression Coefficient when X is Pre - selected

The joint distribution of ε_i is $g(\varepsilon_1 \ldots \varepsilon_N) = (2\pi\sigma^2)^{-N/2} \exp\left[-\Sigma_i \frac{(y_i - \alpha - \beta x_i)^2}{2\sigma^2} \right]$

where ε_i is the value of ε_x when $x = x_i$, $i = 1, 2 \ldots N$. Let y_i is the corresponding value when $x = x_i$. Then, joint density function for y_i is $f(y_1, y_2, \ldots y_N)$ where $f(y_1 \ldots y_N)\, dy_1\, dy_2 \ldots dy_N = g(\varepsilon_1 \ldots\ldots \varepsilon_N)\, d\varepsilon_1 \ldots\ldots d\varepsilon_N$. The y_i are related to ε_i by the equations

$$\alpha + \beta x_i + \varepsilon_i,\quad i - 1, 2, \ldots N$$

and the Jacobian of transformation for ε_i to y_i, for fixed x_i is unity. It follows that

$$f(y_1 \ldots\ldots y_N) = (2\pi\sigma^2)^{-N/2} \exp\left[-\Sigma_i \frac{(y_i - \alpha - \beta x_i)^2}{2\sigma^2} \right]$$

Maximizing $f(y_1 \ldots y_N)$ is equivalent to minimizing $\sum_i (y_i - \alpha - \beta x_i)^2$ which on differentiation with respect to α and β and equating to zero gives

$$\Sigma_i \left(y_i - \hat{\alpha} - \hat{\beta}\, x_i \right)^2 = 0,$$

$$\Sigma_i\, x_i \left(y_i - \hat{\alpha} - \hat{\beta}\, x_i \right)^2 = 0,$$

Simultaneously solving we get $\hat{\alpha} = a$ and $\hat{\beta} = b$ which are the same as obtained by least squares solution. However, note that the above argument does not hold good if X is a random variable. But if the population distribution of X, Y is bivariate normal, then only we get the conditional regression of Y given $X = x$ as linear regression.

We can summarize linear regression by least squares or maximum likelihood methods in reference to ANOVA model as follows :

Source	D.F.	S.S	M.S	F
Linear regression	1	$S_a = b_2 S_X$	$b^2 S_X / 1$	$(N-2)b^2 S_X/(S_Y - b^2 S_X)$
Error	N-2	$S_w = S_Y - b_2 S_X$	$S_w / (N-2)$	
Total	N-1	S_Y	$S_Y / (N-1)$	

If the regression is made between the standardized variables $Z_y = (Y - y)/S_Y$, and $Z_x = (X - \bar{x})/S_X$, and then the linear correlation between X and Y or between Z_y and Z_x is derived through ANOVA model as follows :

Source	D.F.	S.S	M.S	F
Linear regression	1	r^2	r^2	$(N-2)\, r^2/(1-r^2)$
Error	$N-2$	$1-r^2$	$(1-r^2)/(N-2)$	
Total	$N-1$	1	$1/(N-1)$	

The observed F-value can be tested for α-level of significance with population F with 1 d.f. in numerator and N-2 d.f. in denominator. The advantage of linear correlation is that correlation coefficient r is invariant with respect to units of measurements or linear transformations of X, Y. It may be noted that conclusions from $F^1_{N-2,\alpha}$ test is exactly same as that from $\pm\, t_{N-2,\, a/2}$ test on b or r against $H_0 : b$ or $r = 0$ since $F^1_{N-2,\alpha} = t_{N-2,\, a/2}$ numerically.

Functional Relation and Regression between Variables Subject to Error

Often both X and Y whether fixed or random, are subject to experimental error. For random variables experimental error is mixed up with the fluctuation due to the probability distribution. We assume X and Y to be fixed, apart from experimental error, and Y is a linear function of X.

Thus, $X = \xi + \mu$ and $Y = \eta + \nu$ and $\eta = \alpha + \beta\xi$.

ξ and η are true values of X and Y and μ and ν are corresponding errors. Let μ and ν is uncorrelated with each other and for any ξ. Further these are distributed normally with mean zero and variances σ_u^2 and σ_v^2, respectively. The joint density function of X and Y is then,

$$f(x, y) = (2\pi\,\sigma_u\,\sigma_v)^{-1} \exp\,[((x-\xi)^2/2\,\sigma_u^2)-((y-\eta)^2/\,2\,\sigma_v^2)]$$

and $$\text{In}\,L = C - N\log\sigma_u - N\log\sigma_v - 1/2\,\sigma_u^2\,\Sigma_i\,(x_i - x_i)^2$$

$$-\,1/2\,\sigma_v^2\,\Sigma_i\,(y_i - \alpha - \beta\,\xi_i)^2.$$

The right side has $N + 4$ unknown parameters namely α, β, σ_u, σ_v and N values of ξ_i. The MLE's are formed by differentiating with respect to each of these parameters and by equating to zero.

We find $$N\sigma_u^2 = \Sigma\,\beta^2\,\sigma'_u\,(y_i - \alpha - \beta\xi_i)^2/\sigma_v^4$$

or $$\beta^2 = \sigma_v^2\,/\,\sigma_u^2.$$

However, this relation does not hold for all values of u and v since these are random errors.

Therefore, we have to make further assumption that the ratio

$$\sigma_v^2\,/\,\sigma_u^2 = \lambda\ \text{(constant)}.$$

Density $$\sigma_u^2 = \sigma^2.$$

We have then,

$$2N\lambda\sigma^2 = \lambda\Sigma\,(x_i - \xi_i)^2 + \Sigma\,(y_i - \alpha - \beta\xi_i)^2$$

and $$\lambda(x_i - \xi_i) + (y_i - \alpha - \beta\xi_i) = 0$$

We get $$N\alpha + \Sigma_i\,x_i\,\beta = \Sigma_i\,y_i$$

and $$\Sigma_i\,x_i\,.\,\alpha + \Sigma_i\,x_i^2\,\beta = \Sigma_i x_i y_i + (\beta/\lambda)(\Sigma_i\,y_i^2 - \alpha\,\Sigma_i\,y - \beta\,\Sigma_i x_i y_i)$$

Eliminating α from above two equations we get the quadratic in β as

$$s_{xy}\,\beta^2 + (\lambda\,s_x^2 - s_y^2)\,\beta - \lambda\,s_{xy} = 0$$

where s_x^2, s_y^2 and s_{xy} are variances and covariance of the observed sample values x_i, y_i.

The estimator of β is

$$\hat{\beta} = \{s_y^2 - \lambda s_x^2 + [(s_y^2 - \lambda s_x^2)^2 + 4\lambda\, s_{xy}^2]^{1/2}\}/(2 s_{xy})$$

We also get $\sigma^2 = ((N-1)/4N\lambda)\, \{s_y^2 - \lambda\, s_x^2 + [(s_y^2 - \lambda s_x^2)^2 + 4\lambda s_{xy}^2]^{1/2}\}$

Unfortunately this is not a consistent estimator of σ^2 and tends to $\sigma^2/2$ as $N \to \infty$. If we wish to use least squares method, we should weight the square deviation of y_i from $\alpha + \beta x_i$ with a weight inversely proportional to the variance of $(y_i - \alpha - \beta x_i)$.

This variance is $\sigma_v^2 + \beta^2\, \sigma_u^2 = \sigma_v^2 (\lambda + \beta^2)$,

We have to minimize $\Sigma\, (y_i - \alpha - \beta x_i)^2/(\lambda + \beta^2)$.

Differentiating w.r.t. α and β as λ is assumed to be known, we obtain

$$\bar{y} = \hat{\alpha} + \hat{\beta}\, \bar{x}$$

and

$$\hat{\beta}\, \Sigma\, (y_i - \hat{\alpha} - \hat{\beta}\, x_i)^2 + (\lambda + \hat{\beta}^2)\, \Sigma\, (y_i - \hat{\alpha} - \hat{\beta}\, x_i)^2 = 0.$$

Shifting origin to $(\bar{x}, \bar{y})$, $\hat{\alpha} = 0$, we get

$$(\lambda - \hat{\beta}^2)\, \Sigma\, x_i y_i + \hat{\beta}\, \Sigma\, (y_i^2 - \lambda x_i^2) = 0$$

Thus,

$$(\lambda - \hat{\beta}^2)\, s_{xy} + \hat{\beta}(s_y^2 - \lambda\, s_x^2) = 0$$

This is same as MLE but we now get consistent estimator of σ^2 by dividing the S_{min} by the d.f. = (N – 2).

So,

$$\hat{\sigma}^2 = \frac{1}{N-2}\, \Sigma\, \frac{\left(y_i - \hat{\beta} x_i\right)^2}{\lambda + \hat{\beta}^2}$$

$$= \frac{1}{N-2}\, \frac{1}{\hat{\beta}}\, \Sigma x_i \left(\hat{\beta}\, x_i - y_i\right)$$

Therefore,

$$\hat{\sigma}^2 = \frac{N-1}{N-2} \left(s_x^2 - \hat{\beta}\, s_{xy}\right)$$

$$= \frac{N-1}{N-2}\, \frac{1}{\lambda} \left(s_y^2 - \hat{\beta}\, s_{xy}\right)$$

4.7 The Distribution and Hypothesis Test for Pearson Correlation Coefficient

If parent population is uncorrelated and if X and Y are normally distributed then,

$$t = r\,(N-2)^{1/2}\,(1-r^2)^{-1/2}$$

has a student t-distribution with $(N\text{-}2)$ d.f.

Thus, defining $u = r^2/(1-r^2) = t^2/(N-2)$, we easily see that u is a ratio of two independent distributions with 1 and $(N-2)$ d.f respectively. So u has a beta-prime distribution with parameters 1/2 and $(N-2)/2$ that is $f(x) = u^{-12}$ $(1+u)^{-(N-1)/2}/B\left(\frac{1}{2},\frac{N-2}{2}\right)$.

The density formula for r is obtained by equating $2g(r)dr = f(u)du$

We get $g(r) = (1-r^2)^{(N-4)/2}/B\left(\frac{1}{2},\frac{N-2}{2}\right)$ for $N \geq 5$, $g(r)$ is symmetric bell shaped curve ranging between -1 and $+1$.

$$E(r) = 0 \text{ and } E(r^2) = (N-1)^{-1}.$$

The kurtosis is $-6/(N-1)$ which tends to zero for large N, therefore for large N, $g(r)$ is approximately normal. This is true even if X and Y are independent and either X or Y has a normal distribution.

However, if population correlation is not equal to zero, the exact distribution of r is complicated and is given by

$$f_{(r,\rho)} = \frac{N-2}{\pi}\,(1-\rho^2)^{(N-1)/2}\,(1-r^2)^{(N-4)/2}$$

where $$I(\rho r) = \int_0^\infty \frac{du}{(\cosh u-\rho r)^{N-1}}$$

The cumulants are $(n = N-1)$:

$$k_1 = E(r) = \rho - \frac{\rho(1-\rho^2)}{2n}\left[1-\frac{1-9\rho^2}{4n}+.....\right]$$

$$k_2 = V(r) = \frac{(1-\rho^2)^2}{n}\left[1+\frac{11\rho^2}{2n}+...\right]$$

$$\gamma_1 = -\frac{6\rho}{n^{1/2}}\left[1+\frac{77\rho^2-30}{12n}+...\right]$$

$$\gamma_2 = \frac{6}{n}\left(12\rho^2 - 1\right) + \ldots.$$

Thus for $\rho = 0.8$ and $N = 50$ we find $\gamma_1 = -0.71$, $\gamma_2 = 0.82$ which shows that $g(r)$ is far away from Normal. In order to test the hypothesis about r when $H_0 : \rho \neq 0$ and $H_1 : r \neq \rho$, would be complicated. Fisher's z' transformation of r where $= \tanh^{-1} r = 1/2 \ log_e \frac{1+r}{1-r}$ has a variance $1/(N\text{–}3)$ independent of the value of ρ.

Density $\qquad \zeta = \tanh^{-1} r,$

We get, $\qquad E(z') = \zeta + = \frac{\rho}{2n} + O(1/n^2)$ where $n = N\text{–}1$

Also $\qquad V(z') = \frac{1}{n} + (4 - \rho^2)/2n^2 + O(1/n^3)$

$$\gamma_1(z') = \frac{\rho^3}{n^{3/2}} + O(1/n^{5/2})$$

$$\gamma_2(z') = \frac{2}{n} + O(1/n^2)$$

For $\rho = 0.8$ and $N = 50$, $\gamma_1(z') = 0.0015$ and $\gamma_2(z') = 0.042$, which are not significantly different for Normal $\gamma_1 = 0$ and $y_2 = 0$. So z' has a clearly Normal distribution whereas distribution for r is non-normal when population correlation coefficient is non-zero.

4.8 Multiple / Polynomial Regression

Multiple Regression

If the random variable Y (dependent), depends on p other variables X_1, X_2, ... X_p (predictors) then the univariate regression can be exactly determined and in simple regression case but,

$$y_c = \sum_i b_i x_{i,} \; i = 1, 2, 3 \ldots p.$$

To avoid a separate constant term the first variate is denoted X_0 (dummy variable), which is always 1. The b_i are partial regression coefficients and they are estimators of population regression parameters β_i. Individual (a) has moments $x_i \alpha$ ($i = 1, \ldots p$; $\alpha = 1, \ldots N$). The data matrix can be written as

$$\begin{bmatrix} x_{11} & x_{12} & \cdots & x_{1N} \\ x_{21} & x_{22} & \cdots & x_{2N} \\ \cdots & \cdots & \cdots & \cdots \\ x_{p1} & x_{p2} & \cdots & x_{pN} \\ y_1 & y_2 & \cdots & y_p \end{bmatrix}$$

The least square condition can be written as

$$S_k \left(y_k - Z\beta_i \; x_{ik}\right)^2 = \text{minimum}$$

and we get the normal equation :

$$= S_k \Sigma_{j=0}^{p} x_{ik} \, x_{jk} \, \beta_j = S_k \, x_{ij} \, y_k$$

or $$= a^{ii} \, \hat{\beta} = g_i, \; i = 0, 1, 2 \ldots p$$

This is a system of p +1 linear equations on the (p + 1) unknown $\hat{\beta}_j$. In matrix notation it is written as $Ab = g$, whose solution is $b = A^{-1}g$, where $AA^{-1} = I$. As in case of simple linear regression, the ML solution for b is same as the LS solution where b's are linear function of y_k. We know $E(b_i) = \beta_i$ and covariance of b_i and b_j is $\sigma^2 \, a^{ij}$ where a^{ij} is an element of A^{-1}. $V(y_c)$ is given by

$\sigma^2 \sum_{ij} a^{ij} x_i x_j$ and $V_{(y)} = \sigma^2 \left[1 + \sum_{ij} a^{ij} x_i \, x_j\right]$. The unbiased estimator of error variance $\hat{\sigma}^2$ is given by $\hat{\sigma}^2 = \dfrac{1}{N - p - 1} \left[S_i \, y_i - \sum_i b_i \, g_i\right]$.

We can test b_i for statistical significance by t-test as

$$t = \left(b_i - \beta_i\right) \Big/ \left(a^{ij} \, \sigma^2\right)^{1/2}$$

and it has (N–p–1) degrees of freedom.

Therefore, 100 (1–α)% confidence interval for β_i is

$$b_i - \hat{\sigma}\left(a^{ij}\right)^{1/2} t_\alpha < \beta_i < b_i + \hat{\sigma}\left(a^{ij}\right)^{1/2} t_\alpha$$

The variance of the difference between two coefficient b_i and b_j is given by.

$$V(b_i - b_j) = V(b_j) + V(b_i) - 2Cov\,(b_i\, b_j) - \sigma^2\left(a^{ii} + a^{ii} - 2\, a^{ii}\right)$$

Polynomial Regression

Since the predictors X_1 ... Xp are not assumed to be independent, they may be taken as powers of a single variable X_1 as X_1, $X_1^2 = X_2$, $X_1^3 = X_3$, etc. and least square fit would give a polynomial regression of y on powers of X_1. However, we should be careful as to upto, which power of X_1 should be included in the regression equation and other higher powers of X_1 neglected. A rough guide could be the F-test as in ANOVA table. It should be noted that if higher power term is included then its lower power terms couldn't and must not be deleted.

The method of polynomial fit can be substantially improved and computations much reduced if orthogonal polynomials are used. In this method new terms are added independently of the previous terms and tests of significance of coefficients are simplified and independently made. Two polynomials are orthogonal (independent) for the sets of x_α ($\alpha = 1 \ldots N$) if the covariance is zero, i.e., $S_\alpha \left[p_i(x_\alpha).p_2(x_\alpha)\right] = O$.

All polynomials with even subscripts have values that are symmetric about the middle, while those with odd subscripts are skew-symmetric. Therefore, only the lower half of the full table is sufficient. Values for coefficients for orthogonal polynomials have been tabulated for orders up to $N = 104$. Equivalent procedure could be to find principal components of covariance matrix of $X_1 \ldots X_p$ and find multiple regression of Y on the principal components.

When the data are very large, it is used to replace scatter of (x,y) data by grouping these on intervals of y as well as x and make a two-way table. Each column (or x-array) include all the observations with x-values in that class-interval and all these data are assumed to have central value of the interval.

Within one column, the y-values are also grouped in classes and in each class (row) the observations have central value of that y-class. Let's there are p columns and total frequency in i^{th} column is f_i with $\sum_i f_i = N$ (total number of observations).

We then compute arithmetic mean $\bar{y}$ for each column and these y_i are plotted against the central x-values. We get an empirical trend line joining these points. If we fit a least squares straight line to these sets of column means, each weighted by corresponding column frequency, we get the ordinary regression line of y on x.

The sum of square derivatives from column mean is $S_\alpha (y_{i\alpha} - \bar{y}_i)^2$ with ($\alpha = 1, 2 \ldots f_i$). The correlation ratio of y on x denoted as E_{yx} is defined as

$$1 - E^2_{yx} = \sum_{ij} S_\alpha (y_{i\alpha} - \bar{y}_i)^2 / \sum_{ij} S_\alpha (y_{i\alpha} - \bar{y})^2$$

where $\bar{y}$ is over all mean of y.

The denominator can be written as $(N{-}1)\, s^2_y$. This shows that E_{yx} is similar in nature to the Pearson correlation coefficient r. If the regression is linear, then E_{yx} and r are very close to each other, but the more the regression departs from straight line, the more do E_{yx} and r differ.

We can also write $\sum^2_{xy} S_{\bar{y}} / S_y$

where $S_{\bar{y}} = \sum_i f_i (\bar{y} - \bar{y})^2$ and $S_y = \sum_{ij} S_\alpha (y_{i\alpha} - \bar{y})^2 = (N-1)\, S^2_y$.

We may reject linearity of regression if the ratio $\left(E_{yx}^2 - \sigma^2\right)/\left(1 - E_{yx}^2\right)$ is large. If $\bar{y}$ are normally distributed, then this ratio will have *F*-distribution with (*p*-2) degrees of freedom in numerator and (*N*-*p*) degrees of freedom in denominator.

Exponential Regression

Often a variable *Y* increases/decreases with time at a uniform rate such as amount due at compound interest mode or growth of bacteria population.

Mathematically, $dY/Y = k\,dX$, or $Y = A \exp(kX)$

We wish to fit an exponential curve by least squares to a set of *N* pairs of observed value

$$(x_i, y_i), \quad i = 1, 2 \ldots N.$$

We have to calculate A and k by least squares

$$S\left(y_i - A_e^{Kx_i}\right)^2 = \text{minimum}$$

Direct solution by differentiating with respect to *A* and *k* and then equating to zero and simultaneously solving is rather tedious. It is much easier to transform the original nonlinear equation to a linear form as :

$\ln Y = \ln A + kX$

We can easily fit a straight line to this by using weighted least squares to obtain $\ln A$ and *k*. The weighting factor for each observation is made proportional to y_i, so that we minimize

$$S(\ln y_i - \ln A - k\,x_i)\,y_i = \text{minimum.}$$

The normal equations are

$$\ln A \,.\, S\,y_i + k\,S\,(x_i\,y_i) = S\,(y_i\,\ln y_i)$$

$$\ln A \,.\, S(x_i\,y_i) + kS\,(x_i^2, y_i) = S(x_i\,y_i\,\ln y_i)$$

from which $\ln A$ and *k* can be easily computed, by simultanuesly solving the above two equations

4.9 Rank Correlation

Sometimes we rank *N* individuals on basis of some characteristics without actually measuring it (Hardness by Moh's scale). If the same individual is ranked by two judges/two criteria the degree of concordance between two ranking may be of interest. If two judges agree perfectly then i^{th} individual is also i^{th} in second list of rankings.Then the concordance is1; if two judges are diametrically opposite so that i^{th} individual has rank $(N - i + 1)^{th}$ rank in the second list then concordance will be –1. If the ranking are due to pure chance, the concordance should be zero.

Spearman's rank correlation coefficient r_s on the differences $d_i = x_i - y_i$ between the two rankings, and is given by $r_s = 1 - 6 / \sum d_i^2 / [N (N^2 - 1)]$.

Kendall's rank correlation coefficient r_k gives each possible pairs of individuals a score of +1 or –1 if their ranks are of same order or opposite order respectively.

We compute $$r_k = \frac{S}{\binom{N}{2}}$$

where S is total score of the sample.

The same r_k value is obtained in a simple way by writing x_i in their natural order and then for each y_i counting how many numbers x_i there are to the right of this y_i and greater than it. The sum of $\Sigma x_i = P$.

Then, $$r_k = [4P/N(N-1)] - 1$$

The Spearman rank correlation coefficient is same as Pearson's correlation coefficient r if the ranks are considered as actual measured variates.

The rank correlation coefficient can test the association between X and Y, and has the advantage over Pearson correlation coefficient r in that assumptions of normality for at least one of X or Y is not required. We only need that X and Y are independent in the parent population. As N increases, the score S is normal with variance $N (N-1)(2N+5)/18$. Therefore, distribution of r_k is normal. Spearman's r_s also tends to be normal as N increases but more slowly than r_k, with

$$V(r_s) = 1/(N-1).$$

4.10 Distribution - Free Inference

The validity of any statistical inference depends on the input information being correct. For example, in parametric inference based on normal pdf, we assume that the continuous random variable X belongs to the family of Normal density having population parameters μ_0 and σ_0^2, the observations are error-free and independent. However, it is often not possible to prove that the random variable X belong to a normal distribution (with parameters μ_0 and σ_0^2) and therefore it may be simpler if no assumptions regarding the nature of population distribution is made distribution free methods) these tests and inferences may not be as efficient as, efficient as parametric inference (t–, F-tests) but are robust and convenient to compute when the observations are independent and specially when more observations are easily made. The following are important distribution-free or nonparametric tests. These methods are also necessary when observations are not continuous variables but are ranked qualitative variables.

(i) *Sign test*

Let $x_1 \ldots x_n$ be iid observations on a random variable X with continuous d.f. If median is known as as per hypothesis H_0, we get r negative $(X_i - \mu_0)$ and s positive $(x_i - \mu_0)$ where $r + s \leq n$. The individual distribution of r given $(r+s)$ is binomial with $p = 1/2$, and we can use either binomial property tables or chi-square test on 1 d.f., when $(r + s)$ is large. If $(r + s)$ is large (≥ 25) then normal approximation to the binomial may be applied.

(ii) *Two sample tests (Kolmogorov - Smirnov)*

H_0: the two samples come from same population with distribution $F(x)$. We obtain D_{mn} least upper bound $|S_m(x) - S_m(x)|$ where m, n are sample sizes and $S_m(x)$ is observed cumulative relative frequency in sample m.

The asymptotic distribution of D assuming $F(x)$ is continuous is

$$\lim_{m,n \to \infty} P\left(\frac{mn}{m+n} D_{mn} \geq \lambda\right) = L(\lambda) = 2 \sum_{n=1}^{\infty} (-1)^{n+1} e^{2n^2\lambda^2}$$

where for $\lambda = 1.36\ N^{-1/2}$ the $P\ (D_{mn} > \lambda) = 0.05$.

However, if the main interest is to find the difference in mean values if any, then following tests are useful.

(a) *Run test (Wald - Wolfowitz)* : The $m + n$ observations of the two samples are put together and arranged in increasing order. Each observation of first sample put as 0 and send as 1. If $F_1 = F_2$ then we get all possible $[(n + m)!/(m!\ n!)]$ arrangements of m zeros and n unities and hence the distribution of any statistics based on sequence of 0 and 1 can be calculated. Number of zeros and ones is one such statistics. If $F_1 \neq F_2$, the expected numbers of run would be much less than that at a chosen level of significance.

(b) *Signed-ranked test (Wilcoxon)* : This is better than run-test as differences are ranked also. We give ranks for the observation of second sample besides sign obtain in run-test (above).

Let the ranks be $s_1 \ldots s_n$.

If $H_0 : F_1 - F_2$ is true, any selection of n ranks from $1, 2, \ldots (m + n)$ has the same probability $n!\ m!/(n + m)!$ and hence the distribution of sums of ranks can be obtained. If one-sided alternatives (H_1) are considered, a large value of statistics (sum of ranks) indicates departure from H_0. If the alternative (H_1) is two-sided, then both small and large value of statistics are considered for region of rejection. For $m = n$, T statistic is smaller rank sum with

$$\mu = n(n + 1)/4$$

and $\sigma^2 = n(n+1)(2n+1)/24$ for large n (Normal approximation).

(iii) *Chi square tests*

(a) *Simple hypothesis*

Let $n_1, \ldots n_k$ be observed frequencies and $\pi_1, \ldots \pi_k$ be the hypothesis probabilities of a multinomial distribution. To test hypothesis $\pi_i = \pi_i^0$, $i = 1 \ldots k$, we compute χ^2 (Pearson).

$$x^2 = \Sigma \frac{(n_i - n\pi_i)^2}{n\pi_i} = \Sigma \frac{(Obs - Exp)^2}{Exp}$$

This has $(k - 1)$ d.f. as $= \Sigma_{i=1}^{k} \pi_i = 1$.

(b) *Goodness of fit*

Whether cell probabilities can be specified by a fewer unknown parameters $\theta < k$. Then,

$$\chi^2 = \Sigma \frac{(n_i - n\pi_i)^2}{n\bar{\pi}_i} = \Sigma \frac{(Obs - Exp)^2}{Exp}$$

But χ^2 has $(k\text{-}1\text{-}q)$ d.f., where $\hat{\pi}_i = \pi_i(\theta)$. In the above, $\hat{\theta}$ is maximum likelihood estimator of θ with q parameters. This provides overall goodness of fit for all k cells but we can test for any single cell by corresponding χ^2 test (as observed χ^2 is additive) with 1 d.f. If N is large, a test based on normal deviate $N(0,1)$ for deviation in j^{th} cell is

$$\frac{n_j - n\pi_j\left(\hat{\theta}\right)}{\left[\sqrt{n \operatorname{var}\left(\hat{\theta}\right) \pi_i\left(\hat{\theta}\right)}\right]^{1/2}}$$

For a single parameter θ_j, we obtain

$$\operatorname{var}(\theta) = (1 - \pi_j) - 1/\pi_j - 1/\pi_j\, (d\pi_j / d\theta)^2\ 1/i(\theta).$$

where $i(\theta)$ is information per observation.

Fisher's information measure on θ is given by

$$I(\theta) = E\left(\frac{d \log P}{d\theta}\right)^2 = V\left(\frac{d \log P}{d\theta}\right) \text{ as } E\left(\frac{d \log P}{d\theta}\right) = 0 .$$

P is the pdf with parameter θ.

(iv) *Contingency tables*

(a) For large N in a contingency table of r rows and c columns, the independence ($\pi_{ij} = \pi_i . \pi_j$) of rows vs columns can be tested through χ^2 as

$$\chi^2 = \Sigma_{i=1}^{r} \Sigma_{j=1}^{c} \left(\frac{n_{ij} - n..\pi_{ij}}{n..\pi_{ij}} \right)^2 = \Sigma_{ij} \frac{(O-E)^2}{E}$$

This has (r–1) (c–1) d.f. for χ^2 where $n.. = \Sigma_i \Sigma_j n_{ij}$.

The expected cell frequency should be at least 5 for the test to be useful. Some cells are collapsed to obtained $E = 5$. If, however, the marginal cell probability of rows and columns are known, we modify the test

$$\chi_1^2 = \Sigma_i \frac{(n_{i.} - n..\pi_i)^2}{n..\pi_i}, \text{ d.f.} = r - 1$$

$$\chi_2^2 = \Sigma_j \frac{(n_j - n...\pi_j)^2}{n..\pi_{.j}}, \text{ d.f.} = c - 1$$

These represent the discrepancy between the observed and expected marginal frequencies. We can now test whether the observed marginal totals are expected as follows:

$$\chi_3^2 = \chi^2 - \chi_1^2 - \chi_2^2 = \frac{\Sigma\Sigma\left[n_{ij} - n_{i.}\pi._{j.}n..\pi_{i.}\pi._t\right]^2}{n..\pi_{i.}\pi._j}$$

χ_3^2 has (rc – 1) d.f. and can be used to test independence.

(b) If N is small, we should avoid the nuissance (population) parameters $\pi_{i.}$, $\pi_{.j}$ and obtain cell frequencies for hypothesis of independence by computing exact probabilities as :

$$\frac{\prod n_i! \prod n_j!}{n..!} \, |\,|\,|\,| \frac{1}{n_{ij}!}$$

If cell frequencies are not small, we could calculate χ^2 under independent hypothesis and use unit normal deviate exceeding $\sqrt{\chi^2}$ in a 2 × 2 contingency Table as :

$$\chi^2 = n..(n_{11} \times n_{22} - n_1 . \times n._1)^2 / (n_1 . \times n_2 \times n._1 \times n._2)$$

Yates continuity correction of increasing the smaller frequencies by 1/2 but without altering the marginal may be useful to compute χ^2 accurately.

4.11 Contingency Tables (Theory and Application)

There are situations where we cannot measure or even rank individuals on basis of a characteristic but the sample can be divided into a number of classes and number of individuals in each class can be counted. An example would be classifying rock samples on basis of colour (black, green, yellow, white; *X* variable) and grain size (coarse, medium, fine; *Y* variable). We can classify *N* items and get a contingency table as shown, with marginals (MT).

X\Y	C	M	F	MT
B	n_{11}	n_{12}	n_{13}	n_1
G	n_{21}	n_{22}	n_{23}	n_2
Y	n_{31}	n_{32}	n_{33}	n_3
W	n_{41}	n_{42}	n_{43}	n_4
MT	$n.1$	$n.2$	$n.3$	$n.. = N$

Here, $N = \sum_i n_i = \sum_i n_j$ and relative frequency of ij^{th} cell $f_{ij} = n_{ij}/N$ is approximately close to π_{ij} population probability.

If *i* and *j* are independent we know that,

$$\pi_{ij} = \pi_i . \pi_j . \text{ where } \sum \pi_{ij} = 1.$$

The degree of associatiation between *X* and *Y* in the population, designated by

$$\phi^2 = \sum_{ij} (\pi_{ij}{}^2 / \pi_i \pi_j) - 1.$$

$\phi^2/\min(s, t)$ varies from 0 to 1 and behaves like r^2.

Here, s = number of row classes and t = number of column classes.

The expected frequency in the ij^{th} cell is $N\,\pi_{ij}$; and

$$\chi_x^2 = \sum_{i,j} \frac{(f_{ij} - N\pi_{ij})^2}{N\pi_{ij}}$$

$$= \sum_{i,j} \frac{(f_{ij} - N\pi_i\pi_j)^2}{N\pi_i\pi_j}$$

π_i and π_j are estimated as $n_i./N$ and $n_j./N$ respectively and these are MLEs χ_s^2 has $(s-1)(t-1)$ degrees of freedom. However, if the expected frequency are small (say, less than 5) we can estimate ϕ^2 by f^2 as $f^2 = \sum_{ij}\left(\frac{n_{ij}^2}{n_i.n_j}\right) - 1 = \chi_s^2/N$.

Here, $\frac{f^2}{\min(s,t)}$ is a measure of association in the sample.

Coefficient of the contingency is given by

$$C = \left[f^2/(\min(s,t)-1)\right]^{1/2}$$

and a low value indicates little or no association.

4.12 Non Parametric Tests for Randomness and Robustness

For testing randomness and robustness by parametric method we require that the sample values be independent. However, if we measure n consecutive maximum temperature (X) for n days at a place, temperature on i^{th} day and on $(i + 2)^{th}$ day X_i and X_{i+1} are positively correlated and are not independent. Similarly, the number of motor accidents may be strongly and positively correlated for X and X. Therefore, serial correlation should be more effective tool for checking randomness in cyclical or time series type data and one cannot use ordinary sample correlation coefficient (r). We can compute serial correlation coefficient as

$$r_1 = \sum_{i=1}^{n} \frac{(x_i - \bar{x})}{s_x} \frac{(x_{i+1} - \bar{x})}{s_x}$$

If x_i are homogeneously regular then r_i is asymptotically normal with $E(r_1)$ and $V(r_1)$is given by

$$E(r_1) = \left(s_1^2 - s_2\right)/(n-1)$$

and $$V(r_1) = \frac{s_2^2 - s_4}{n-1} + \frac{s_1^4 - 4s_1^2 s_2 + s_1 s_3 + s_2^2 - 2s_4}{(n-1)(n-2)} - E^2(r_1)$$

where $$s_k = x_k^1 + \ldots\ldots x_n^k.$$

Thus, a t - test can now be applied as $t = [(r_1 - E(r_1))/\sqrt{V(r_1)}\,]$. If the value is acceptable for the normal hypothesis at .05 level then the sample data are sequentially independent (random), otherwise the sequence is dependent (non-random).

It is well known that parametric tests are valid if the population distribution is known to belong to a specific probability density function on which parametric estimation and hypothesis tests are based. However often we do not know the nature of density for the population and hence the estimation of parameters and tests of hypotheses would be suspects.

Many tests based on sample means are quite robust because of Central Limit Theorem for large sample sizes, which induces Normal density for the $\bar{x}$ based on n samples. However t-test based on difference of two means may not be robust unless variances of the two variables are also equal. Rank-sum test for two samples also suffers from the same problem.

4.13 Solved Examples

Example 1

Let the population density be non-normal.

$$f(x) = x\,e^{-x},\ x > 0.$$

We test if n samples has mean μ_0 and variance σ_0^2 assuming the population to be Normal.

We know, for large n, $E\bar{X} = \mu$ and $V(\bar{X}) = \sigma^2/n$ substitute s^2 for σ^2 (as n is fairly large). Thus, σ^2 is treated as known in a t-test for mean μ. We have $\mu_0 : \mu = \mu_0$ and $\mu_1 : \mu \neq \mu_0$ and hence, acceptance region for μ_0 is given by

$$= |\bar{x} - \mu_0| < t_{n-1,0.5}\ s/\sqrt{n\mu_o}$$

which is true for both whether population density is either Normal (H_0) or xe^{-x}, $x > 0$ (H_1), so long as n is very large. Therefore, t-test for mean is robust. But test for variance is not robust. Here, $H_0 : \sigma^2 = \sigma_0^2$ against $H_1 : \sigma^2 \neq \sigma_0^2$ and under Normality we have, $\chi_1^2 < ns^2/\sigma^2 < \chi_2^2$. If n is very large we can treat ns^2/σ^2 to be Normally distributed and

$$E\,(ns^2/\sigma^2) = (n-1),\ V(ns^2/\sigma^2) = 2(n-1).$$

So acceptance region for large sample would be

$$\left|\frac{ns^2}{\sigma_0^2} - (n-1)\right| < t_{n-1,\,.05}\ \sqrt{2(n-1)}$$

However, if parent population is $f(x) = xe^x$, $x > 0$, we can get,

$$V(s^2) = \frac{\mu_4 - \mu_2^2}{n} - \frac{2(\mu_4 - 2\mu_2^2)}{n^2} + \frac{\mu_4 - 3\mu_2^2}{n^3}$$

which for large n reduces to $V(s^2) = \dfrac{\mu_4 - \mu_2^2}{n}$

Thus, $E(ns^2/\sigma^2) = n\dfrac{\mu_4 - \mu_2^2}{\mu_2^2}$ and $\mu_2 = 2$ and $\mu_4 = 24$.

So, $V(ns^2/\sigma^2)$ for this population density = $5n$, whereas for normal density we know,

$$E(s^2) = \frac{n-1}{n}\sigma^2 \text{ or } E(ns^2/\sigma^2) = (n-1).$$

The acceptance region under H_1 is

$$\left|\left(\frac{ns^2}{\sigma_0^2} - (n-1)\right)\right| < t_{n-1,.05}\sqrt{5n} \cong 1.58\left(t_{n-1,.0.5}\ s/\sqrt{n}\right)$$

Therefore, serious error is induced if population density is really not normal and hence we accept H_0 under Normal density.

For large n, ns^2/σ^2 may be treated as normal and we get a t–test as

$$t = \sqrt{n}\left(s^2 - \left(\frac{n-1}{n}\right)\sigma^2\right)/\left(\mu_4 - \mu_2^2\right)^{1/2}$$

with sample values replacing population values.

Example 2 : (Independence of colour vs size of igneous rocks)

Suppose we classify 100 igneous rocks according to their colours (light, medium, dark) and grain size (coarse, medium, fine) to test the independence or otherwise. The resulting observed contingency table is given below along with the numbers, which are expected under the independence hypotheses $\chi^2_{obs} = 103$, and tabular χ^2 at 5% level is 9.5 and at 1% level is 13.3 for four d.f. ($\chi_{4:05} = 9.488$; $\chi_{4:01} = 13.277$).

Size\colour	Light	Medium	Dark	Marginal total
Coarse	26 (10)	2 (7)	2 (13)	30
Medium	3 (7)	15 (4)	2 (9)	20
Fine	5 (17)	5 (11)	40 (22)	50
Marginal total	34	22	44	100

Therefore, we conclude that in igneous rocks, colours and sizes are not independent but are strongly associated. Geologically, this means that light coloured rocks tend to be coarse-grained and dark coloured rocks have high probability of being fine-grained. This inference seems geologically and mineralogically reasonable as low temperature minerals like quartz and feldspars are usually coarse-grained and light coloured whereas high temperature minerals like pyroxene and amphiboles are usually dark colored and tend to be fine grained (especially in volcanic rocks). On the other hand, if the observed χ^2 would have been non-significant, that is, lees than $\chi_{4;.05}$(= 9.488) then we would have accepted that colours of rocks are independent of their grain size χ^2_{obs} = 25.6 + 3.6 + 9 + 2.3 + 30.3 + 5.4 + 8.4 + 3.3 + 15 = 103.

Note that rocks tend to be either light-or dark-coloured as well as coarse or fine- grained and less frequent in the category medium-coloured or medium grained.This is due to mode of occurrence of igneous rocks that are either volcanics(fine-grained due to fast rate of cooling and dark coloured due to basaltic composition) or plutonic (coarse-grained due to slow rate of cooling and light colored due to granitic composition).

Example 3

The friction (*y*) in a machine increases with load (*n*) on the machine. Eight experimental data for (x_i, y_i) are as follows : (make a simple linear regression for *y* over *x*).

x	23.4	44.7	65.7	86.8	107.8	128.8	149.6	171.0
y	3.4	4.7	5.4	6.8	7.5	8.8	9.6	11.0

Using the first and last two data we get

$$\bar{x}_i = 34.05,\ \bar{y}_i = 4.05$$

$$\bar{x}_3 = 160.3,\ \bar{y}_3 = 10.3;$$

Also, we have $\bar{x} = 97.15,\ \bar{y} = 7.15.$

Therefore, $\hat{\beta} = (\bar{y}_3 - \bar{y}_1)/(\bar{x}_3 - \bar{x}_1) = .0495$ and $\hat{y} = .0495x + 2.34.$

Also, we compute (8-3) s_x^2 = 226.8 + 2224.0 + 229.0 = 2679.8

$$5\,s_y^2 = 7.8525;\ 5\,s_{xy} = 143.25.$$

So, $S\left(\hat{\beta}\right) = \beta_y^2 - 2\beta s_{xy} - \beta s_x^2 = .0354.$

We get, $t = (\bar{x}_3 - \bar{x}_1)\,(\hat{\beta} - \beta)/[2S(\beta)/d.f.]^{1/2}$ from which 95% confidence interval for $\hat{\beta}$ is calculated as .0437 to .0529.

Example 4

The correlation coefficient between scores obtained by 20 students on two tests is 0.65. What are the 95% confidence limits for population correlation coefficient ρ ?

Answer :

z' is normally distributed. So, 95% confidence limits will be z' $\pm$ 1.96/ $\sqrt{17}$.

Since $E(z') = \zeta + \rho/2n = \zeta + \frac{\rho}{38}$, the limits for ξ will be

$z' - \rho/38 \pm 1.96/\sqrt{\pm 17}$

But $z' = \tanh^{-1} 0.65 = 0.775$

Therefore, ξ varies from (.775 – .017 $\pm$ .475 =) .283 to 1.233.

The corresponding limits for ρ (tanh [1] ξ) are .276 to .844.

Example 5

Separate samples of sizes 50,70and 100 gave correlation coefficients .72, .680, .77. What is the value of *r* to use as an average?

Answer :

z' are .908, .829 and 1.020.

The weighted mean for z' is

$$\bar{z}' = \frac{1}{211}[47(.908) + 67(.829) + 97(1.020)] = .934.$$

The corresponding *r* is 0.733. Arithmetic average of r values is incorrect since population correlation is non-zero.

Example 6

A drug to prevent air sickness was tested on 25 men given drugs and another 25 men without drugs. Both groups were tested on a rocking machine and result were as follows :

	Sick	Not sick	Marginal Totals
With drug	10	15	25
Without drug	19	6	25
Marginal Totals	29	21	50

$$\chi^2 = \frac{50.(60-285)^2}{25.25.29.21} = 6.65$$

So, the drug was an effective prevention for air sickness.

References and Further Reading

1. Anderson, T. W., 1984(3rd Edn.) An Introduction to Multivariate Statistical Analysis, NewYork, Wiley.
2. Cramer, H., 1946, Mathematical Methods of Statistics, Princeton University Press.
3. Cochran, W.G. and Cox, G.M., 1954, Experimental Design, NewYork, Wiley, p577.
4. Keeping, E. S., 1962, Introduction to Statistical Inference, Affiliated East-West Press, NewDelhi.
5. Krumbein, W. C., 1965, An Introduction to Statistical Models in Geology, NewYork, McGrawHill, New York.
6. Plackett, R. L., 1960, Principles of Regression Analysis, Oxford, Clarendon Press.
7. Rao, C. R., 1973, Linear Statistical Inferences and Applications, NewYork, Wiley, p625.
8. Siegel, S., 1956, Non-parametric Statistics for the Behavioral Sciences, New York, McGrawHill.
9. Walsh, J. E., 1962, Handbook of Non-parametric Statistics, Van Nostrand, London.

5 Multivariate Analysis

5.1 Introduction

It is statistics for analyzing multiply correlated measurements made in one or more samples. Strictly speaking multiple regression/correlation, polynomial regression, partial correlations are univariate statistical procedures (involving one r.v.) but these are included here since same matrix procedures form the basis of inference.

If X is a p-variate random vector we have $p(p-1)/2$ covariances/correlations in addition to p means and p variances. The mathematical model is multivariate normal distribution (mnd), which is preferably defined by characterization that all linear compounds are mnd since multivariate density function may or may not exist. On basis of sets of variables and sets of populations studied, multivariate methods can be classified into four cells as follows :

	One Population	More than one population
One set of variables	Principal component Factor Cluster	Multivariate analysis of variance (MANOVA) Discriminant function(DF) Classification function(CF)
More than one set of variables	Multiple regression Polynomial regression Partial correlation Canonical correlation	Multivariate analysis covariance (MANCOVA)

The random vector measured on geological samples (individuals) from a homogeneous population is often correlated but not necessarily mnd. This problem may be approximately solved by normalizing the marginal distributions of the random vector. By characterization theorem,if the random vector is mnd, then every linear compounds [marginals, principal components and conditional (regressed)distribution] are mnd.

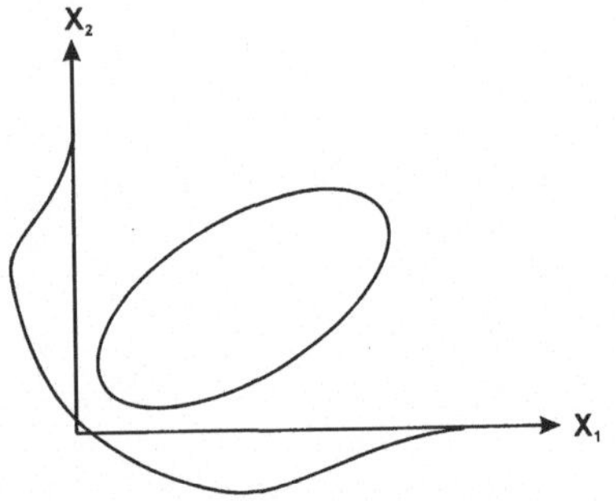

(a) MND with marginal distribution

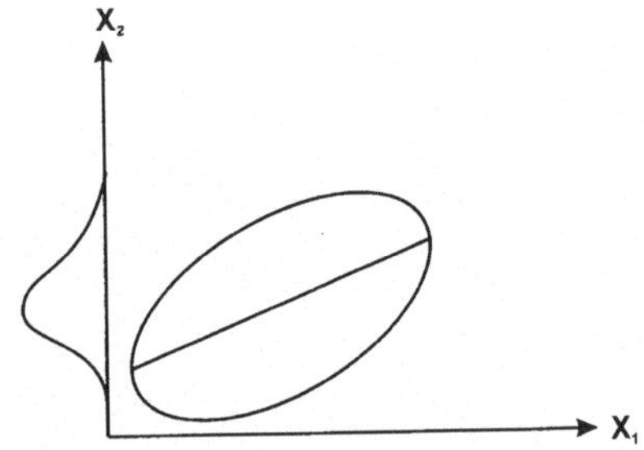

(b) MND with conditional distribution

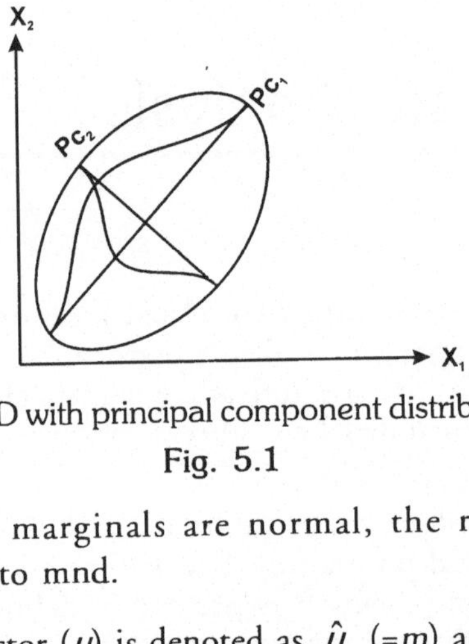

(c) MND with principal component distribution

Fig. 5.1

We assume that if marginals are normal, the random vector (after normalization) belongs to mnd.

The mle of mean vector (μ) is denoted as $\hat{\mu}$ (=m) and is given by

$$m = \hat{\mu} = \bar{x} = 1/\mathrm{N} \sum_{i=1}^{N} x_i$$

and its dispersion (covariance) matrix as

$$D = \text{mle } \Sigma = 1/N \sum_{i=1}^{N} (x_i - m)\,(x_i - m)'$$

or, $$D = \text{mle } \Sigma = 1/N(x_i - \bar{x}_i) - mm' - mm'.$$

The combined estimator is $D_u = \Sigma_u = N/(N-1)\ D$.

The non-diagonal covariances (correlations) r_{jk} can be computed as

$$r_{jk} = d_{jk} / \sqrt{d_{jj}\, d_{kk}}$$

where d_{jk} represent the jk^{th} component of D matrix.

It may be noted that if X is mnd then m is distributed independent of D.

The multivariate data reduction is as follows :

Mean vector $\quad m = 1/N \sum_{i=1}^{N} x_i$ and

Dispersion (covariance) matrix $D = 1/N \sum_{i=1}^{N} (x_i - m)(x_i - m)'$

Transforming x_i to the standardized vector $z_i = (x_i - \bar{x})/\sqrt{S_i^2}$

we get $D = R = 1/N \sum_{i=1}^{N} z_i z_i'$ where correlation coefficient r_{jk} of R is given by

$$1/N \sum_{i=1}^{N} z_i z_i'$$

Let v be the coefficient of a component of z as $y_i = v' z_i$

The standard deviation of y is

$$S_y = \sqrt{v' D_z v}$$

Dividing out S_y and v we get $c = v/S_{y.}$ A standardized component (factor) with zero mean and unit variance is $f_i = c' z_{i.}$

The correlation of f_i with elements of the vector variable z is given as

$$r_{zf} = 1/N \sum_{i=1}^{N} z_i f_i = 1/N \sum z_i (c' z_i)' = 1/N \sum z_i z_i' c = R_z c$$

which is called factor structure.

We can test the null hypothesis for equality (homogeneity) of mean vectors $H_2 : \mu = \mu_0$, assuming dispersion D is homogeneous, by Hotelling T^2 as

$$T^2 = N (m - \mu)' D_\mu^{-1} (m - \mu).$$

Under $H_2 : \mu = \mu_0$, $(N - p)(T^2/(N - 1))$ has F distribution with p and N–p degrees of freedom. This is a simple extension of univariate t-test.

If T^2 is smaller than F_p, N–p, α, then H_2 may be accepted, otherwise it is rejected. This is an univariate extension to t-test for multivariate populations as regards equality of mean vectors given those covariance matrices are homogeneous.

5.2 Matrices

A matrix is an arrangement of pq elements belonging to any field in p rows and q columns. Matrices are represented by capital letter (say, A) or as (a_{ij}) where a_{ij} is the ij^{th} element of matrix with $i = 1 \ldots p$, $j = 1 \ldots q$.

The following operations are possible :

Addition : $A + B = B + A$ or $A + (B + C) = (A + B) + C = A + B + C$

Scalar multiplication :

$$cA = (ca_{ij}),\ (c + d)\ A = cA + dA$$
$$c(A + B) = cA + cB.$$

Matrix multiplication AB is defined only when the number of columns in A is same as number of rows in B.

If so, $AB = (c_{ij}) = C$

where, $c_{ij} = \Sigma\ r\ (a_{ir}\ b_{rj})$.

In general matrix multiplication is not commutative, i.e. $AB \neq BA$. In general associative law of multiplication is valid i.e. $A(BC) = (AB)C = ABC$. Null matrix is 0 with all its elements 0. A square matrix is called unit or identity matrix (I), if its diagonal elements are unity and non diagonal elements are all zero.

We have, $A + 0 = A$, $IA = AI = A$

when products are defined. Therefore *0* and *I* behave as zero and unit elements of the number system.

Transpose of matrix A (written as either A^T or A' is found by interchanging the rows and columns. Thus, ij^{th} entry of A' is $(j, i)^{th}$ the entry of A.

So, $(AB)' = B'\ A'$ and $(ABC)' = C'\ B'\ A'$.

A vector in R^n, written usually as a column, is a matrix of order $n \times 1$. If X and Y are two column vectors, their inner product (X, Y) can be written $X'\ Y$. If A is $m \times n$ matrix, then Ax is a column vector, which is a linear combination of columns vector of A.

A^* is said to be a conjugate transpose of A, if ij^{th} entry of A is the complex conjugate of the ji^{th} entry of A^*.

Thus, $(AB)^* = B^*\ A^*$, $(ABC)^* = C^*B^*A^*$

Rank of matrix is the number of independent rows (= number of independent columns) in the matrix.

A square matrix A of order $(m \times m)$ is nonsingular if its rank is m. In such a case we have a unique matrix A^{-1} such that

$$AA^{-1} = A^{-1}\ A = I.$$

If A^{-1}, B^{-1} exist then,

$$(AB)^{-1} = B^{-1} A^{-1}$$

and $$(ABC)^{-1} = C^{-1} B^{-1} A^{-1}.$$

We can also easily check that $(A')^{-1} = (A^{-1})'$

or $$(A*)^{-1} = (A^{-1})*$$

If B is a $m \times n$ matrix and $BC = I_m$, then C is called right regular inverse of B written as B_R^{-1}.Similarly B_L^{-1} may exist if

$$B_L^{-1} = C \text{ and } CB = I_n$$

A right inverse exists if $r(B) = m \leq n$ and a left inverse exists if $r(B) = n \leq m$.

A square matrix A of complex elements is said to unitary if $A^* A = I$.

In such case, $A^* = A^{-1}$ and $AA^* = I$.

When the elements of the matrix are real, the matrix is said to be orthoganal if

$$A' A = A A' = I$$

A matrix can be partitioned as $A = \begin{pmatrix} P & Q \\ R & S \end{pmatrix}$ where P, Q have same number of rows and P, R have same number of columns.

Thus, $$A^T = \begin{pmatrix} P' & Q' \\ R' & S' \end{pmatrix} \text{ and}$$

$$AB = \begin{pmatrix} P & Q \\ R & S \end{pmatrix}\begin{pmatrix} E \\ G \end{pmatrix} = \begin{pmatrix} PE & + & QG \\ RE & + & SG \end{pmatrix}$$

$$AC = \begin{pmatrix} P & Q \\ R & S \end{pmatrix}\begin{pmatrix} E & F \\ G & H \end{pmatrix} = \begin{pmatrix} PE + QG & PF + QH \\ RE + SG & RF + SH \end{pmatrix}$$

provided these products exist.

A square matrix A is symmetric if $A = A^T$ and hermitian if $A = A^*$

Matrix can be reduced to simpler forms through elementary operations on their columns/rows. These operations preserve the rank of the matrix as these operations are equivalent to multiplication by nonsingular matrices.

Type of Matrix	Definition
Symmetric	A = A'
Hermitian	A = A*
Idempotent	$A^2 = A$
Positive definite (p.d.)	XAX > 0 for non null X
Positive semi definite (p.s.d.)	XAX >= 0 for non null X
Non negative definite (nnd)	X* AX >= 0 for all X
Normal	AA* = A*A
Orthogonal	AA'=A'A=I
Unitary	AA* = A*A = I
g-Inverse	A^-

Determinant of a square matrix $A = (a_{ij})$ of order m is a real valued function of elements a_{ij} as

$$|A| = \Sigma\ (\pm)\ a_{1i},\ a_{2j} \cdots a_{mp}$$

where summation is taken over all permutations $(i, j \ldots p)$ of $(1, 2 \ldots m)$ with a plus sign.

If $(i, j, \ldots p)$ is an even permutation and minus sign if it is an odd permutation. Cofactor of a_{ij} is denoted as $A_{ij} = (-1)^{i+j}$ (determinant of A excluding i^{th} row and j^{th} column). A matrix X can be linearly transformed to another matrix Y through $Y_{(m)} = AX_{(n)}$ where A is $m \times n$ matrix of real elements. The transformation of X to Y is one to one only when A is non-singular (A^{-1} is unique and exists).

Then, $X = A^{-1}\ Y.$

A linear transformation $Y = CX$ is orthogonal if $CC' = I$.

Since $|C|^2 = 1$, the transformation is non-singular. The condition $CC' = I$ implies that row vectors are orthogonal and of unit length forming an orthonormal basis in R^n.

Since $|C| = \pm 1$, $C^T = C^{-1}$ and therefore, $C^T C = I$ which implies that columns also from an orthonormal basis.Orthogonal transformation preserves distances as well as angles.

If A^- is g-Inverse (generalized inverse of A)of matrix A then, $AA^-A = A$. A^- is not necessarily unique.

$$A^-A = H \text{ with } H^2 = H \text{ (idempotent)}.$$

If $\quad A^- = A$ we have, $|A| = r\,(A) = r\,(H) = tr\,(H)$.

Also if A^- exists, then $r\,(A^-) \geq r\,(A)$.

5.3 Matrix Computation

A quadratic form plays an important role in statistical rank. It is defined as

$$Q = \Sigma_{i=1}^{n} \Sigma_{j=1}^{n} a_{ij}\ x_i\ x_j = X'AX$$

where X is the column vector of variables and A = $[(a_{ij} + a_{ji})/2$ which is symmetric,is called the matrix of quadratic form. A quadratic form is positive definite if $X^TAX > 0$ for all non-null X and is negative definite $-XAX$ is positive definite. It is positive semi definite(psd) if $X^TAX \geq 0$ and $= 0$ for at least one nonnull X and nsd if $-X^TAX$ is psd. Matrix A is pd/nd when the associated quadratic form is p.d/n.d.

Considering a non-singular linear transformation $Y=B^{-1}X$ or $X=BY$, we obtain

$$Q = X^T AX \rightarrow Y^T B^T ABY = Y^T CY$$

which is a quadratic form in the new variables with $C = B^TAB$. Under non-singular linear transformation we have the definiteness of Q is invariant.

Every Q can be reduced to a form containing square terms only i.e. $p+q = r$ where p is the number of positive diagonal elements, q is number of negative diagonal elements and r is number of nonzero elements in the reduced matrix.

If Q is p.d, $q = 0$, hence $p = r$.

Therefore, if $X = BY \quad X^TAX \rightarrow Y^TB^T ABY = Y^T Y \Rightarrow B^T AB = I$

or $A = CC^T$, where $C^T = B^{-1}$.

Every p.d. matrix A can be expressed as CC^T.

A necessary and sufficient condition for matrix A to be p.d. is $|A| > 0$.

For a real symmetric matrix A of order m we have the following characteristics:

(i) If $r(A) = r$, then equation $|A-\lambda I| = 0$ has zero as a root of multiplicity $m-r$.

(ii) All the latent roots are real and latent roots are chosen to be real

(iii) All the latent vectors are orthogonal i.e $P_i^T\ P_j = 0$.

(iv) If A is p.d all roots are positive; if A is nnd all roots are nonnegative.

(v) If X is non-null vector, there exists a latent vector $Y \in$ linear manifold $M(X, AX, A^2 X\ldots)$.

If A is a real symmetric of order $m \times m$, then there exists an orthogonal matrix P such that

$$P^T AP = \Lambda \text{ or } A = P\ \Lambda P^T$$

where Λ is the diagonal matrix of latent roots.

We can write $A = P\Lambda\ P^T = \lambda_1\ P_1\ P_1^T + \ldots + \lambda m P_m\ P_m^T$

which is known as the spectral decomposition of a matrix A.

If A and B be real symmetric m × m matrices of which B is p.d, then there exists, a matrix R such that $A = (R^{-1})' R^{-1}$ and $B = (R^{-1}) R^{-1}$ where Λ is a diagonal matrix. In other words there exists a transformation $Y = R^{-1} X$ such that the two quadratic forms $X^T AX$ and $X^T BX$ are simultaneously reduced to simpler forms containing square terms only i.e.

$$X^T AX \rightarrow \lambda_1 y_1^2 + ... + \lambda_m y_m^2$$

$$X^T BX \rightarrow y_1^2 + ... + y_m^2$$

5.4 Homogeneity of Covarience / Correlation Matrices

Box (1949) has given a test criterion M for homogeneity of several covariance matrices $\Delta_k = \Delta$ for $k = 1, 2, \ldots g$.

$$M = (N-2) \log_e | D_w | - \Sigma_{k=1}^{g} (N_k - 1) \log_e |D_k|$$

D_k is dispersion for k^{th} group and D_w is the pooled within group dispersion, we compute following :

$$A_1 = \left(\Sigma_{k=1}^{g} \frac{1}{N_k - 1} - \frac{1}{N - g} \right) \frac{2p^2 - 3p - 1}{6(g-1)(p+1)}$$

$$A_2 = \left(\Sigma_{k=1}^{g} \frac{1}{(N_k - 1)^2} - \frac{1}{(N - g)^2} \right) \frac{(p-1)(p+1)}{6(g-1)}$$

If is positive, then we calculate

$$n_1 = (g-1)\, p(p+1)/2,$$

$$n_2 = (n_1 + 2)/(A_2 - A_1^2)$$

and $$b = n_1/(1 - A_1 - (n_1/n_2)).$$

We then test H_1(homogeneity of covariance matrices) as

$$F_{n_1}^{n_2} = M/b$$

If observed F is greater than tabular F at α level then H_1 is rejected.

If $(A_2 - A_1^2)$ is negative, then we calculate

$$F_{n_1}^{n_2} = n_2 M/n_1 (b - M)$$

where $$n_1 = (g-1)\, p(p+1)/2,\ n_2 = (n_1 + 2)/(A_1^2 - A_2)$$

and $$b = n_2/(1 - A_1 + (2/n_2)),$$

If observed F is greater than tabular F at a level then H_1 is rejected.

Homogeneity of several correlation matrices can be tested in a similar way using

$$\Delta_k = \Delta \quad \text{for } k = 1, 2, \ldots g$$

where Δ's are now correlation matrix.

We use R_k instead of D_k and R_w is now used as pooled within group correlation matrix (in place of D_w). Although these tests are fairly robust for departures from mnd, one has to be careful.It should be noted that tests for differences in mean vectors such as Hotelling's T^2 or linear discriminants are useful provided covariance(correlation) matrices are homogeneous. Hence, H_1 test is to be accepted before proceeding to test hypothesis H_2 for homogeneity of mean vectors of several groups. If some variances in the covariance matrices are much larger than others, homogeneity in covariance matrices can be brought about by making scale change in those input components of the multivariate vector.This makes the variances nearly homogeneous and hopefully the covariances of groups can then be found to be approximately homogeneous. Details of multivariate statistical methods are included in the next three chapters along with geological examples and case studies.

5.5 Examples

Example 1

Compute approximate mean and covariance for the transformation $y(\underline{\underline{x}})$ where $\underline{\underline{x}} = (x_1, \ldots x_i, \ldots x_p)^T$ with known mean vector $\underline{\underline{\mu}}$ and covariance matrix $\underline{\underline{V_{ij}}}$, but pdf of $\underline{\underline{x}}$ unknown and x_i are not necessarily uncorrelated.

Answer :

Taylor expansion of $\underline{\underline{x}}$ as $y(\underline{\underline{x}}) \approx y(\underline{\underline{\mu}}) + \Sigma_{i=1}^{p} \left[\frac{\partial y}{\partial x_i}\right]_{x=\mu} (x_i - \mu_i)$

$$E\left[y(\underline{\underline{x}})\right] \sim y(\underline{\underline{\mu}})$$

since $\quad E(x_i - \mu_i) = 0$ for all i.

$$E\left[y^2(\underline{\underline{x}})\right] \approx y^2(\underline{\underline{\mu}}) + \Sigma_{i,j=1}^{p} \left[\frac{\partial y}{\partial x_i}\frac{\partial y}{\partial x_j}\right]_{x=\mu} V_{ij}$$

So, $$\sigma_y^2 \approx \Sigma_{i,j=1}^{p} \left[\frac{\partial y_k}{\partial x_i}\frac{\partial y}{\partial x_j}\right]_{x=\mu} V_{ij}$$

For a set of m functions $y_1(\underline{x}), \ldots y_m(\underline{x})$ the covariance matrix is

$$U_{kl} = \mathrm{Cov}\,[y_k, y_l] \approx \Sigma_{j=1}^{p} \left[\frac{\partial y_k}{\partial x_i} \frac{\partial y_l}{\partial x_j} \right]_{x=\mu} V_{ij}$$

or $$U = AVA', \text{ where } A_{ij} = \left[\frac{\partial y_i}{\partial x_j} \right]_{x=\mu}$$

This will be simplified if x_i are not correlated, that is if $V_{ij} = \sigma_1^2$ and $V_{ij} = 0$ for $i \neq j$ and then we get,

$$\sigma_y^2 \approx \Sigma_{i=1}^{p} \left[\frac{\partial y}{\partial x_i} \right]^2 \Bigg|_{x=\mu} \sigma_i^2$$

and $$U_{kl} \approx \Sigma_{i=1}^{p} \left[\frac{\partial y_k}{\partial x_i} \frac{\partial y_l}{\partial x_j} \right]_{x=\mu} \sigma_i^2.$$

Example 2 (Targetting of mineral deposits)

The formation of a mineral deposit is controlled by a number of factors. Complex processes operate in nature to produce equally complex products and both are involved with a large number of factors which necessitates a probabilistic approach. Multivariate statistical procedures, fuzzy set theory, Bayesian statistics, etc. are some of the methods used to delineate areas where the prospect of locating the mineral deposit is considerably higher than the adjacent regions. This greatly reduces the exploration costs confining detailed exploration efforts to such higher potential areas. Multivariate statistical analyses like factor analyses, logistic regression etc., Boolean logic, fuzzy logic and Bayesian methods can be used for the aforesaid purpose but here we discuss in some detail the last one of the methods.

Bayesian Methods

These methods of data integration use a probability framework, using the prior probability of occurrence of mineral deposit and combine it with some evidences to calculate the posterior probability of deposit occurrence in the area. The mineral deposits are considered as point objects and the goal is to predict the presence of such a set of point objects. These are treated as binary, being either present or absent. The model requires that each be treated as a small area object occurring within a small unit cell.

The probability of finding ore prospect given the presence of at least one ore body can be expressed by the conditional probability :

$$P\{D|B\} = P(D \cap B)/P(B)\}$$

Further, in order to obtain an expression for the posterior probability of a prospect in terms of the prior probability and a multiplication factor, the following calculations are made :

The conditional probability of being on a binary map B, given the presence of a deposit is defined as :

$$P\{B|D\} = \{P(B \cap D)/P(D)\}$$

Combining these two equations, we get

$$P\{D|B\} = P\{D\}.\{P(B|D)/P(B)\} \qquad .....(a)$$

A similar expression can be derived for the posterior probability of a deposit occurring given the absence of a pattern as follows:

$$P\{D|\overline{B}\} = P\{D\}.\{P(B|D)/P(\overline{B})\} \qquad .....(b)$$

The weights of evidence method uses the odds form of the probability.

Dividing both sides of (a) by $P(\overline{D}/B)$,

$$P\{D|B\}/P(\overline{D}|B) = P\{D\}\{P(B|D)/P(\overline{D}|B)P(B)\} \qquad .....(c)$$

Now, by definition

$$P\{\overline{D}|B\} = \{P(\overline{D} \cap B)/P(B)\} = \{P(B|\overline{D})\,P(\overline{D})/P(B)\}$$

Substituting in (c) and rearranging,

$$P\{D|B\}/P(\overline{D}|B) = P(D)/P(\overline{D})P(B)P(B|D)/P(B|\overline{D}) \qquad .....(d)$$

The odds of a deposit $O(D)$are equal to $\{P(D)/(1-P(D))\}$, or

$P\{P(D)/P(\overline{D})\}$.

Substituting in (d) and simplifying,

$$O(D|B) = O(D)\{P(B|D)/P(B|\overline{D})\} \qquad .....(e)$$

where $O(D/B)$ is the conditional (posterior) odds of D given B, $O(D)$ is the prior odds of D and $\{P(B/D)/P(B|\overline{D})\}$ is known as the sufficiency ratio LS. The natural logarithm of both sides of (e) are taken and In (LS) is the positive weight of evidence, W^+, which is calculated from the data. Then,

$$\text{logit}(D/B) = \text{logit}\{D\} + W^+ \qquad .....(f)$$

A similar expression can be derived for the odds of the conditional probability of D given the absence of the binary indicator pattern, with the result being :

$$O(D|\overline{B}) = O(D)\{P(\overline{B}|D)/P(\overline{B}|\overline{D})\}$$

The term $\{P(\overline{B}|D)/P(\overline{B}|\overline{D})\}$ is called the necessity ratio, LN, and the negative weight of the evidence, W^- is given by ln(LN). The corresponding logit expression is :

$$\text{logit } \{D|\overline{B}\} = \text{logit (D)} + W^- \qquad .....(g)$$

LN and LS are also called likelihood ratios.

When evidence from several maps are combined, the weights are calculated from each map independently and then combined in a single equation. This requires an assumption of conditional independence.

Let P(D) be the prior probability of the occurrence of the mineral deposit.

The conditional probability of a deposit occurring, given the presence of two predictive map patterns B_1 and B_2 is :

$$P\{D|(B_1 \cap B_2)\} = P(D \cap B_1 \cap B_2)/P(B_1 \cap B_2)$$

which can be written as

$$P(D|B_1 \cap B_2)$$

$$= \{P(B_1 \cap B_2|D)P(D)\}/[\{P(B_1 \cap B_2|D)P(D)\} + P(B_1 \cap B_2|\overline{D})P(\overline{D})] \qquad .....(h)$$

If two patterns are independent with respect to a set of deposit points, then,

$$P(B_1 \cap B_2|D) = P(B_1|D)P(B_2|D)$$

Substituting in (h) and simplifying,

$$P\{D|(B_1 \cap B_2)\} = \text{logit } (D) + W_1^+ + W_1^-$$

where the subscripts 1 and 2 refers to the weights determined independently for maps 1 and 2.

The general expression for combining i = 1, 2, ... n is :

$$\text{logit } (B_1 \cap B_2 \cap .....\cap B_N) = \text{logit } (D) + \Sigma W_2^+$$

In this formula, W^- should be used instead of W^+ if the ith map pattern is absent instead of being present.

5.6 Exercises

1. *What are the main multivariate factors you could consider for exploration of sedimentary exhalative (sedex) deposits ?*

Answer :

(a) host rock with syntectonic sedimentation as submarine debris flows and large differential subsidences.

(b) geochemical haloes such as Mn,Fe,Mg in limestones

(c) Ba,K,Na,Zn,B in pelites.

(d) presence of paleobrines to prevent oxidation of sea floor sulfides(i.e. at low paleolatitudes)

2. *Which are the main multivariate factors for the exploration of porphyry type copper sulfide ore deposits ?*

Answer :

Surface indications for porphyry deposits are :

(a) gossan with porphyry boxworks

(b) halo of muscovite

(c) halo of sericite and argillites (chlorites)

(d) map of paleo-volcanic crust with calc-alkaline rocks above subduction zones.

3. *Which are the main multivariates for the exploration of epithermal deposits ?*

Answer :

Vertical and inclined diamond drilling to decipher :

(a) volcanic history and stratigraphy

(b) patterns of hydrothermal alteration

(c) calc-alkaline volcanism above subduction zones

(d) presence of artesian conditions for deposition of Mississippi Valley type sulfide deposits (epigenetic)

4. *What are the different multivariate factors for 'ore targeting' around a known are deposit ?*

Answer :

(a) study the known deposits and identify its genesis (ore type).

(b) use the multivariate factors for that ore type for search around 20 to 40km radius of the known deposit.

(c) additional factors include remote sensing data, different geophysical and geochemical maps, paleogeologic maps for metallogenic epochs.

5. *What are the multivariate factors which help in exploration and identification of diamondiferous kimberlite pipes and lamproites ?*

Answer :

Kimberlitic and lamproitic magmas should be generated at higher-pressure (stable) field for diamond and these magmas should come up fast enough so that diamond to graphite transformation do not take place. Keeping this genesis we can use the following factors :

(a) presence of chromite and

(b) presence of perovskites

(c) pyrope with high Na^{+} content

(d) diopside with high Cr^{+++} content

A linear discriminant function with normalized variable inputs such as log(pyrope fraction), log(diopside fraction), log(Na^{+} fraction in pyrope) and log(Cr^{+++} fraction in diopside) could be useful to discriminate between diamondiferous and non-diamondiferous kimberlites/lamproites. Diamondiferous rocks should also have chromites and/or perovskites.

6. *What are the multivariate factors for the deposition of uranium ores in sedimentary Environments ?*

Answer :

(a) presence of uranium ores in source areas/source rocks

(b) change from oxidising to reducing environment which may be due to water table, eutrophic layer in lakes/seas, lateral facies changes inducing more organic matter and clays etc.

(c) climate changes providing cyclic source/deposition

(d) presence of host rocks like sandstones, conglomerates, basal unconformity surfaces etc.

7. *During 1950's and 1960's W.C.Krumbein(1965) and his colleagues concentrated on Trend Surface Analysis and predictions of geological variables in 1D, 2D, 3D space using this method. It is in fact polynomial regression using geographical (spatial) coordinates as predictors. Critically examine this methodology from a statistical viewpoint and list some of its drawbacks.*

Hint :

(i) Since X, Y, Z are assumed Normally distributed, higher powers such as x^2, x^3, xy^2, y^3 etc. cannot be Normal. So, multiple (polynomial) regression is not always valid.

(ii) Sample points may be dependent on neighbours, so that multivariate statistical models may not be valid. But time series or random field methodology should be more appropriate which does not require that samples are independent.

(iii) However, if sample points are widely spaced, then these could be independent (uncorrelated) and trend surface may be useful in such cases if corresponding F-tests are highly significant for the low order polynomials thought to be geologically meaningful.

References and Further Reading

1. Agterberg, F. P., Bonham-Carter,G.F., Cheng,Q. and Wright, D. F., 1993, Weights of evidence modelling and weighted logistic regression for mineral potential mapping.
In : Davis, J.C. and Herzfeld (Eds), Computers in Geology-25 years of progress, Oxford University Press, 13-22.

2. An, P., Moon,W. M. and Rencz,A., 1991, Application of fuzzy set theory for integration of geological, geophysical and remotesensing data, Can. J. Expl. Geoph., 27, 1-11

3. Anderson, T. W., 1984, An introduction to multivariate statistical analysis, New York, Wiley.

4. Bonham-Carter, G. F., 1994, Geographic Information Systems for geoscientists, Pergamon Press, Oxford, p397.

5. Box, G. E. P., 1949, A general distribution theory for a class of likelihood criteria : Biometrika, 36:317-346.

6. Burrough, P. A. and R.A.McDonnel, 1998, Principles of Geographical Information Systems, Oxford Uni. Press, Oxford, U.K., p333.

7. Cooley, W. W. and Lohnes,P.R., 1971,Multivariate Data Analysis, New York, Wiley, p364.

8. Krumbein, W.C.,1965, An Introduction to Statistical Models in Geology, New York, McGrawHill.

9. Rao, C. R., 1973, Linear statistical inference and its applications, New York, Wiley, p625

10. Sahu, B. K., 1991, Stochastic models for sedimentary uranium ores for optimal mining and resource estimation, IAEA Shortcourse at Hyderabad, AMD.

11. Sahu, B. K., 1995, Statistical Inference from geochemical and petrographic data, Proc. Recent Res. Geol. W. India, p59-64.

12. Sahu, B. K., 1998, Philosophy and strategies for optimal Exploration of Mineral Deposits, SGAT bulletin : 1(1) : 2-14.

13. Sahu, B. K., 2001, Modelling Sedimentary ore Deposits.
In : Proc. 3rd Indian Conf. On Computer Applications in Mineral Industry, Oxford and IBH Publishing Co., New Delhi, p135-139

14. Sahu, B. K. and Balasubramanian, L., 1997, Multilevel data integration in mineral exploration - A GIS based approach, Geos., IIT, Kharagpur, 13(1) : 20-29.

15. Zimmermann, H. J., 1985, Fuzzy set theory and its applications : Kluwer-Nijhoff Publishing, Boston, p363.

6 Principal Component and Factor Analysis

6.1 Introduction

The methods discussed in this chapter are useful for studies on one population (group). Principal components are linear combinations of original variables/ observations belonging to mnd, so that the variance of each component, that is orthogonal to other principal components, is maximum. The main idea is to transform the original vector space to a principal component space where the principal component scores are uncorrelated , i.e., the covariance (correlation) matrix of mnd is linearly transformed such that it becomes an orthogonal matrix of variances (as all covariances among the orthogonal principal components are zero). Mathematically, a real symmetric matrix is diagonalized where the diagonal elements are eigenvalues corresponding to which the vectors directions are called eigenvectors (Fig. 6.1).

In Factor Analysis each (orthogonal) principal components is standardized (mean= zero, variance=unity) so that each corresponding factor now is of equal importance (variance). Thus, the diagonal/covariance matrix of principal components is converted to an identity matrix. We can also find the factor structure of covariance (correlation) matrix by using oblique factors, instead of

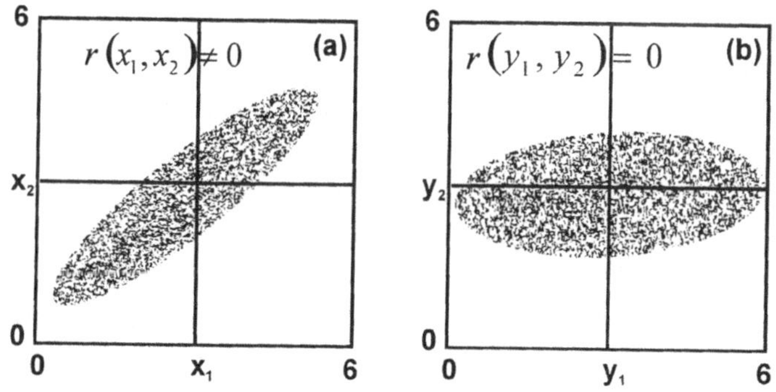

Fig. 6.1 Correlation between x_1 and x_2 is non-zero but the same data referred to principal components y_1 and y_2 has correlation $r(y_1, y_2) = 0$

orthogonal factors, if necessary (Fig. 6.2 & 6.3). But oblique factors are correlated with each other as well as with measurement vectors, so these are much more difficult to interpret (compared to orthogonal factors), unless such oblique factors have very simple structure.

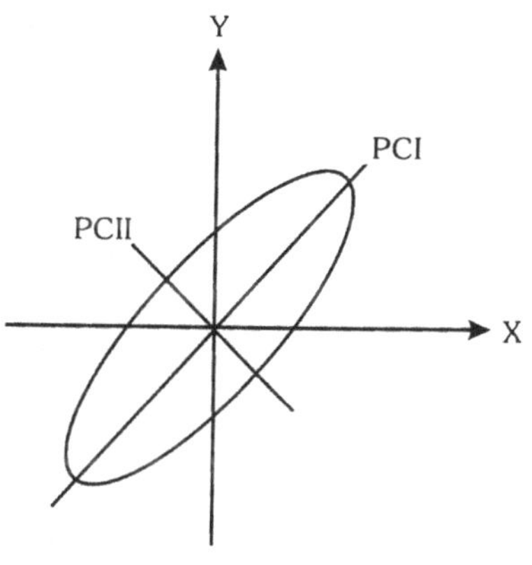

Fig. 6.2 Principal Components (I and II) for bivariate MND

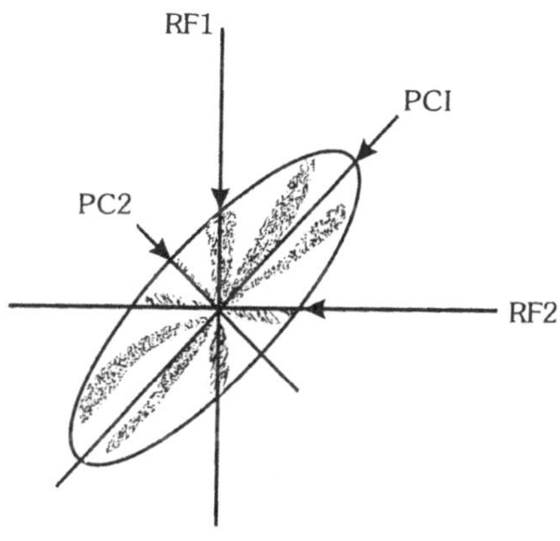

Fig. 6.3 Rotated factors (RF1, RF2) gives simple structure compared to PC1, PC2 systems

In cluster analysis, the aim is to find natural clusters among samples, which are more homogeneous than others. This is an empirical and nonparametric approach using either distance functions or correlation coefficients. Although, these techniques can sometimes give useful insights, the logic of including only significant variables for clustering is not always very clear. Hence, completely unrelated objects may belong to same cluster on basis of some irrelevant measurements made on them. Therefore, we do not emphasize clustering methods.

6.2 Eigenstructure of a Real Symmetric Matrix

The input matrix is either covariance matrix (D) or correlation matrix (R) for a single population.This is a real symmetric matrix with non-zero non-diagonal elements which indicate that the components of multivariate measurement vectors are correlated. Our aim is to diagonalize this matrix such that the diagonal elements (variances or eigenvalues λ) are maximized. Corresponding to each diagonal element we can compute a direction (eigenvector) and these directions are mutually orthogonal (hence uncorrelated). Occasionally, the number of nonzero eigenvalues (m) may be less than rank of input matrix (p) which simplifies our inference due to reduction of the multivariate space._However, most often the number of non-zero eigenvalues (m) will be equal to rank of input matrix ($p=m$), but still as correlations are all zero in the eigenspace inferences on basis of eigen vectors are much easier.

A factor is a standardized principal component (eigen vector) having mean zero and variance equal to unity. In factor analysis, we exploit the dominance of a few (significant) eigen values (m) as compared to rest (p-m) eigenvalues which are statistically insignificant and can be neglected. This gives a reduction in multivariate_space from p to $m < p$, and simultaneously induces all correlations among the significant eigenvectors to be zero. However, the numbers of significant eigenvectors retained for further analysis are to be carefully determined and the remaining residual eigen vectors dropped from the study are considered as noise. Identification of retained significant components (factors) with respect to the observed multivariate input vector is to be solved, as well.

The computation of eigen structure of a real symmetric matrix from one population such as D or R is shown below :

The full eigenstructure of R is given by

$$RV = V\Lambda$$

where $V'V = VV' = \mathrm{I}$ and $V'RV = \Lambda = D_y$.

We can get back $\mathrm{R} = V\Lambda V'$

Here,

R = correlation input matrix,

V = matrix of all eigenvectors

Λ = L matrix of eigenvalues (of R) or of D_y (dispersion matrix of transformed vector y).

However, before finding eigenstructure of R, we must test whether $R = I$ in which case the input vector is already an eigenvector and no components analysis is required. We use the test as follows (Bartlett 1950).

$$\chi_{\frac{1}{2}p(p-1)} = -\left[(N-1) - \frac{1}{6}(2p+5)\right] \log_e |R|$$

and if observed rank is greater than $\chi_{\frac{1}{2}p(p-1),.05}$, .05 only then component analysis is required.

From matrix theory (spectral decomposition) we have

$$R = \lambda_1 v_1 v_1' + \lambda_2 v_2 v_2' + \lambda_p v_p v_p' = \sum_{j=1}^{p} R_j$$

If m components are significant and rest $(p\text{-}m)$ components are noise we have

$$\hat{R} = R_1 + \ldots.. R_m,\ \tilde{R} = R_{m+1} + \ldots.. R_p$$

so that $= R = \hat{R} + R \cong \hat{R}$ with rank $m < p$.

In actual practice drawing of eigenvectors and eigenvalues is made much faster by powering the input matrix R to a high value say R^{64}. This makes the first eigenvalue λ_1^{64} dominate over the rest and eigenvector v_1 converges quickly. We compute residual matrix $\tilde{R} = R - \lambda v_1 v_1'$ and then power it to $\tilde{R}_1^{64}$ as input to draw λ_2 and v_2 and so on, till exhaustion of input matrix R (i.e., extraction of all eigenvalues and eigenvectors). We also know from matrix theory that :

$$\prod_{i=1}^{p} \lambda_i = |D_y| \text{ or } |R| \text{ and } \sum_{i=1}^{p} \lambda_i = tr\,Dy \text{ or } trR$$

The principal component y_j is standardized to yield principal factor

$$f_j = \left(y_j / \sqrt{\lambda_j}\right)$$

Therefore, $f = \Lambda^{-\frac{1}{2}} y$

but as $y_j = v_j' z$ we get,

$$f_j = v_j' z / \sqrt{\lambda_j}.$$

Since $y = v' z$ we have, $f = \Lambda^{-\frac{1}{2}} v' z = B'z$ where B = factor score coefficients $= v\Lambda^{-\frac{1}{2}}$.

Factor structure $S = V\Lambda V^{-\frac{1}{2}} = V\Lambda^{-\frac{1}{2}}$; Predicted $\hat{R} = SS'$,

Residual $\tilde{R} = R - SS'$.

To determine the number of significant eigenvectors (m) is very important for scientific and applied research. If sample size N, is large, then $\tilde{R}$ tends to I and correlations at the non-diagonal position in should all be nonsignificant at the .05 level of $\chi^2_{\frac{1}{2}p(p-1)}$. However, in case of small sample sizes, over extraction of eigenvectors should be strictly avoided through χ^2 - test for $|\tilde{R}|$ tends to zero.

We compute

$$\chi^2_{\frac{1}{2}(p-m)(p-m-1)} = -\left[(N-1) - \frac{1}{6}(2p+5) - \frac{2}{3}n\right]$$

$$\log_e \left[|R| / \left\{\prod_{j=1}^{m} \lambda_j \left(p - \sum_{j=1}^{m} \lambda_j\right) / (p-m)\right\}^{p-m}\right]$$

which is tested for significance at 0.05 level.

Another useful test is to plot λ_j versus j or standard deviation of cumulative eigenvalues $\sum_{j=1}^{r=p} \lambda_j$ computed on independent replicate samples of size N for the same population versus j. The position of inflexion or minima subject to $\sum_{j=1}^{r \leq m}$ includes major part of information (Sahu 1973) gives a unique value of m for retaining the significant numbers of eigenvectors (eigenvalues). $\sum_{j=1}^{m} \lambda_j$ trace D_y (or trace) gives the total proportion of variance explored by m principal components (factors) retained. Sahu(1973) points out that the two main problems in factor model are :

(i) Determination of common factor space (m) or equivalently how much error (D) can be expected and neglected ; and

(ii) How to interpret the retained m factors that are not uniquely fixed in the common factor space. Sahu (1973) introduced a *second order criterion* , of minimization of variance of cumulative eigenvalue in subpopulations (or in population having the same factors and similar factor structures) acting in conjunction with the *first order criterion* that the average residual variance should not exceed 15 percent of total variance (Rao, 1965), to obtain the common factor space (m)

uniquely. In case the second order criterion fails to give any decision, a larger number of samples should be studied in each subpopulation (or population) concerned so that the correlation matrices are stabilized. The retained factors are not unique in the common factor space and therefore rotation to an optimal desired position will be very useful for identification of factors, test of hypotheses, etc.

The elements of eigenvectors are called factor loadings(v_{jk}) which is the correlation of factor f_j with variable z_k. Sum of squares of columns of $S = V\Lambda^{+1/2}$ gives the proportion of variance extracted by a factor $\sum_{J=1}^{M}\left(\Sigma_{K=1}^{P} s_{kj}^2/p\right)$ indicating the importance of factor j in factor analysis. Sum of squares of rows of S gives communality $h_k^2 = \Sigma_{j=1}^{m} s_{kj}^2$, which indicates the importance of variable K in factor analysis. Further j can be identified in terms of variables K on which it is strongly loaded i.e., $|\vartheta_{jk}|$ is close to 1 and is independent of variables K where $|\vartheta_{jk}|$ is zero. Similarly, all the factors j = 1 to m can be identified.

We can form either the correlation matrix of variables, which is R-mode analysis by summing over N samples, otherwise correlation matrix of samples that is Q-mode analysis by summing over p variables. Both R and Q mode matrices have the same set of eigenvalues. But often p is much smaller than sample size N, therefore we prefer (for less computations) the R-mode correlation matrix (order $p << N$) to be the input matrix for components and factor analysis. As the eigenvalues of the set are same whether R-mode or Q-mode analysis was performed, the final inferences would be identical.

Often the loadings in matrix V between factor and variables are not close to unity or close to zero, hence we are in doubt whether the variable is loaded in the factor which serves to identify the factor. Therefore, in practice, orthogonal rotation of retained m factors is necessary to identify the factors as well as for inference on factors and for geological hypothesis testing etc .The main aim of orthogonal rotation to simple structure is that factor loadings matrix should have elements which are either 0 or ± 1 such that the variance of the squared loadings of factor matrix is maximized (Kaiser 1958).

That is, normal varimax is given by

$$V = \sum_{k=1}^{m} V_k = \sum_{k-1}^{m} \left\{ \frac{\left(\Sigma p \Sigma_{j=1}^{P} \dfrac{s_j^2 k}{h^2{}_j}\right)^2 - \left(\Sigma_{j=1}^{P} s_{jk}^2 / h_j^2\right)^2}{p^2} \right\}_{max}$$

where h_j^2 is communality of variable *j*. This orthogonal rotation (called Varimax Rotation) is done on a computer by step by step small iterative rotations of two by two factors to maximum until all retained m factors are exhausted to reach the final optimal position of simple structure (i.e. maximum variance of loading matrix).

The rotated eigenvalues associated with *j* th rotated factor is λ'_j, and it can be computed as (Sahu, 1973 Eqn. 3.1) :

$$= \lambda'_j = \sum_{i=1}^{p} b_{ij}^2$$

where b_{ij}s are the rotated loadings of jth factors on the variables. Clearly, Griffith's procedure (1966) of judging the importance of rotated factors on the basis of unrotated (principal component) eigenvalues is not statistically valid. However, rotation does not affect communalities (h_i^2) of the variables which remain constant for any variable.

In m factor model, y is the p-dimensional mnd r.v. with structure

$$y = V\hat{F} + G$$

where $\hat{F}' = (F_1 \ldots F_m)$ with each $F_i \sim N_i(0, 1)$

and $G' = (G_1, \ldots G_p)$ with each $G_i \sim N_i(0, \delta_i)$.

F' s are common factors (signals) and *G*'s are error factors (noises) which are independent of all *F*' s , i.e. r (*F*, *G*) = 0.

Hence $D_y = VV' + \Delta_1$

where Δ is diagonal with elements $\delta_1 \ldots \delta_p$ in diagonal positions.

If we rotate orthogonally the common factors $\hat{F}$ to $\hat{F}^*$ positions to reach varimax structure, then

$$y = V^* \hat{F}^* + G^*$$

where $V^* = VT$ (T is orthogonal rotation matrix , i.e. $T'T = TT' = I$)

Then dispersion matrix

$$D_Y = V^* V^{*\prime} + D_2 = VTT'V + \Delta_2 = VV' + \Delta_2$$

Hence, we obtain $\Delta_1 = \Delta_2$

Thus, orthogonal rotations of factors do not change the communalities/ error variances of the input variables as well as the total variance explained by *m* common factors. However, as new loadings are close to either 0 or $\pm$ 1, the rotated factors are easily explained (identified) through the input variables on which they are strongly loaded. It should be noted that the total explained variance by m common factors is compatible with the measurement error variances of input variables (error variances of samples in Q–mode analysis). Factor scores should be calculated on basis of m rotated common factors as

$$S^* = V^* \Lambda^{1/2}$$

where,

$$\Lambda^* = (I - \Delta_2)$$

6.3 Factor Model Applied to Sphericity and Roundness Data

Factor analysis is a useful exploratory tool in multivariate statistical analysis and it can be applied to discover and interpret relations among variables (samples) or to test hypotheses. Factor analysis reduces the dimensions of the multivariate system to a smaller number, called the common factor space (m), and this helps in clearer understanding and interpretations. Although not essential to factor analysis, the raw data are standardized and normalized for better statistical studies later on such as regression or discrimination. R-mode analysis extracts the relations between the variables whereas Q-mode analysis extracts clusters of samples. Recently a new technique of interpretation has been proposed (Sahu 1973) applicable to R-mode orthogonal factor models and this paper aims at giving the details of steps for such interpretations using the sphericity and roundness data as example. The computations are made on IBM-1620 computer following Sampson's (1968) programme.

Patro (1970) has studied the sphericity and roundness of quartz grains in three common sieve fractions (A.S.T.M. –45 + 50; –50 + 60; –60 + 70) from river, beach and aeolian dune environments. Fifty grain sphericity and grain roundness were measured in each sample in each sample from any sieve fraction and subpopulation sizes were 32,38 and 30 respectively. Sahu and Patro (1970) have shown that ($\Psi/(1 - \Psi)$) and $-\log(P)$ normalize grain sphericity and grain roundness data, respectively, where and P is Wadell roundness = $2\bar{r}/b$. Mean, standard deviation, skewness and kurtosis were computed by the standard moment method for grain sphericity and grain roundness data of each sample from each sieve fraction.

Since the size difference in the above sieve fractions are not great , we may assume that the factor structure is similar and thus all three sieve fractions may be added for factor analysis purposes to yield sample sizes 96, 104, and 90. These fairly large sample sizes produce stable 8 × 8 correlation matrices for each environment. The result of principal component analysis of these correlation matrices are included in Table 1 as none of the eigenvalues are zero, straight application of first order criterion yields the component rather than the factor model. Study of Table 6.1 indicates that one minima occurs at factor 2 and the other at factor 6 and the corresponding average residual variances are, respectively, 57.47% and 9.95%. Thus, according to second order criterion (Sahu, 1973) there are six common factors in this system.

Table 6.1 Principal Component eigenvalues and cumulative eigenvalue percents in three environments

A. River Environment

Number	Eigenvalues	Cum. Eigenvalue Percent
1.	1.7568682	21.9608
2.	1.6182848	42.1894
3.	1.3659367	59.2639
4.	1.1012843	73.0300
5.	0.7898976	82.9037
6.	0.5702389	90.0317
7.	0.5543864	96.9615
8.	0.2430716	9.9999

B. Beach Environment

Number	Eigenvalues	Cum. Eigenvalue Percent
1	2.0113347	25.1416
2	1.4633458	43.4335
3	1.3089034	59.7948
4	1.0628436	73.0803
5	0.7775886	82.8002
6	0.5998155	90.2978
7	0.5076891	96.6440
8	0.2684750	99.9999

C. Aeolian Environment

Number	Eigenvalues	Cum. Eigenvalue Percent
1	1.8199522	22.7494
2	1.5384893	41.9805
3	1.2997206	58.2270
4	1.0358358	71.1749
5	0.8566270	81.8828
6	0.6344997	89.8140
7	0.5046210	96.1218
8	0.3102513	99.9999

Table 6.2 Computation of second order criterion for obtaining the common factor space (m)

	Range of Cum. Eigenvalue%	Variance of Cum. Eigenvalue%	Average residual Variance
1.	3.1808	1.8291	76.7161
2.	1.4530	0.1420	57.4655
3.	1.5678	0.4239	40.9048
4.	1.9054	0.7874	27.5716
5.	1.0209	0.0714	17.4711
6.	(min.) 0.4838	(min.) 0.0391	(min.) 9.9522
7.	0.8397	0.1198	3.4242

Note : Number of common factors is six.

The six rotated common factors were varimax rotated (Kaiser, 1958) in the common factor space to yield rotated loadings that are easily interpretable. Table 3 and 4, respectively, give the data for principal component and varimax rotated solutions. Table 5 presents the rotated factors in their order of importance in different environments as suggested by the associated rotated eigenvalues. Table 6 lists the mean values of the variables in different environments. Interpretation of these results should be based on rotated factors, rotated factor loadings, rotated eigenvalues etc. together with the data of Table 6.3. In the fine sand size range sphericity changes are negligible as chipping and fracturing of grains are very difficult. But changes in roundness can be considerable and this doesn't necessarily change the sphericity of the grains. Thus, it is reasonable to hypothesize that in the fine sand size range roundness factors reflect abrasion processes whereas sphericity factors reflect processes of selective sorting on the basis of grain sphericity. Hydraulic conditions of the environments also affect the interpretation of sphericity and roundness factors.

Table 6.3

A. River Environment

Factor → Variable ↓	1	2	3	4	5	6
SM	0.0518	–0.4445	0.1078	0.7631	–0.2445	0.0556
SSD	–0.1154	0.6438	–0.1274	0.3184	0.5558	0.2688
SSK	0.1233	0.5489	0.5402	0.1630	–0.4075	0.3684
SK	–0.2801	–0.5781	–0.4974	0.1988	0.0743	0.3821
RM	0.2768	0.5100	–0.6340	0.3096	–0.2252	–0.1825
RSD	0.4818	–0.2620	0.5522	0.3018	0.4011	–0.1622
RSK	–0.8821	–0.0068	0.2933	–0.1073	–0.0182	0.0788
RK	0.7486	–0.2046	–0.0775	–0.3913	0.0063	0.3839
Eigenvalue	1.7569	1.6183	1.3660	1.1013	0.7899	0.5702

Table 6.3 *Contd...*

B. Beach Environment

SM	–0.0232	0.7166	0.3156	–0.3178	1.0628	–0.5285
SSD	0.3352	0.3511	–0.2707	0.6051	0.5269	0.0433
SSK	0.5002	–0.0336	0.3433	–0.4033	0.2021	0.0748
SK	–0.6117	0.3635	–0.3685	–0.0203	–0.3335	0.1904
RM	–0.5454	–0.1945	0.5948	0.4426	0.0347	–0.0781
RSD	0.2980	0.6824	0.3789	–0.1316	0.0746	0.5021
RSK	0.7791	0.0013	–0.4235	0.0155	–0.2702	–0.1360
RK	–0.5298	–0.5298	–0.4291	-0.4679	0.5177	0.0095
Eigenvalue	2.0113	1.4633	1.3089	0.0098	0.7776	0.5998

C. Aeolian Dune Environment

Factor → Variable ↓	1	2	3	4	5	6
SSD	0.3352	0.3511	–0.2707	0.6051	0.5269	0.0433
SM	–0.0291	0.2398	0.5931	–0.1228	0.7522	0.0857
SSD	0.4227	–0.4824	–0.4457	–0.0387	0.2298	0.5688
SSK	–0.7246	0.2809	–0.1336	0.3100	0.0811	0.1133
SK	0.5438	–0.3575	0.4436	–0.3368	–0.1820	–0.1675
RM	0.2944	0.7700	–0.0558	–0.3877	–0.1420	0.2060
RSD	–0.1631	–0.2518	0.7170	0.3498	–0.3205	0.3733
RSK	–0.5691	–0.6202	–0.0978	–0.3267	0.1782	–0.1496
RK	0.6185	–0.0193	–0.0884	0.6560	0.2219	–0.2419
Eigenvalue	1.8200	1.5385	1.2997	1.0358	0.8566	0.6345

Table 6.4 Rotated factor loadings and eigenvalues with six factors for each environment.

Factor (rotated) → Variable ↓	1	2	3	4	5	6
A. River Environment						
SM	–0.0232	0.7166	0.3156	–0.3178	1.0628	–0.5285
SM	–0.1277	0.0726	–0.0806	0.8584	–0.1527	–0.2621
SSD	–0.1292	0.0895	–0.1420	–0.1003	0.9533	–0.0205
SSK	0.0166	0.9614	0.0136	0.0188	0.1063	0.0053
SK	0.1160	–0.5258	0.1792	0.6121	0.0145	0.3812
RM	0.0027	0.0408	–0.9165	0.0081	0.1730	0.2155
RSD	0.1588	0.0357	0.1428	0.1407	0.0284	–0.9035
RSK	–0.6308	0.0639	0.6065	–0.0319	0.0406	0.3308
RK	0.9396	0.0126	0.0092	–0.0846	–0.1197	–0.0641
Rotated Eigenvalue	1.3527	1.2212	1.2872	1.1500	0.9903	1.1907
Rotated Factor	RNN	SNN	RM-RSK	SM	SSD	RSD

Table 6.4 *Contd...*

B. Beach Environment

SM	–0.0624	0.9759	0.0105	0.0114	0.0040	0.1938
SSD	–0.0258	–0.0118	–0.1132	0.9652	–0.0441	0.0855
SSK	0.8349	–0.0877	–0.1132	–0.2361	0.0133	0.1147
SK	–0.8399	–0.0097	0.0310	–0.1967	0.1977	0.0754
RM	–0.0323	–0.0209	0.9101	–0.0175	–0.2003	–0.1479
RSD	0.0319	0.2037	–0.0467	0.0884	–0.1241	0.9515
RSK	0.1414	–0.0440	–0.8272	0.1580	–0.3648	–0.1170
RK	–0.1354	–0.0002	0.0239	–0.0393	0.9578	–0.1291
Rotated Eigenvalue	1.4474	1.0042	1.5402	1.0608	1.1472	1.0213
Rotated Factor	SNN	SM	RM-RSK	SSD	RNN	RSD

C. Aeolian Dune Environment

SM	0.0203	0.0440	0.0570	–0.0014	0.9932	–0.0797
SSD	0.1347	–0.0764	–0.0756	0.0872	–0.0840	0.9713
SSK	–0.8057	–0.0845	0.1059	–0.1483	0.0017	–0.2052
SK	0.8774	0.0231	–0.1521	–0.0048	0.0241	–0.0263
RM	0.0385	0.8915	–0.2323	–0.1928	0.0703	–0.0560
RSD	0.0496	–0.0971	0.9725	–0.0362	0.0596	–0.0763
RSK	–0.0301	–0.7796	–0.1417	–0.4973	0.0175	0.0507
RK	0.1128	0.0089	–0.0555	0.9502	0.0003	0.0994
Rotated Eigenvalue	1.4551	1.4275	1.0662	1.2183	1.0029	1.0140
Rotated Factor	SNN	RM-RSK	RSD	RNN	SM	SSD

Note: *RNN: Roundness nonnormality factor*
SNN: Sphericity nonnormality factor

Table 6.5 Relative importance of sphericity and roundness factors (Rotated Factors) in each environment and over the three environments.

Relative rank of factors → **Variable ↓**	6	5	4	3	2	1
River	RNN	RM-RSK	SNN	RSD	SM	SSD
Beach	RM-RSK	SNN	RNN	SSD	RSD	SM
Dune	SNN	RM-RSK	RNN	RSD	SSD	SM
Total rank over	RM-RSK	SNN	RNN	RSD	SSD	SM
Environments	16>	15>	14>	8>	6>	4

Note: *Higher number means greater importance according to rotated eigenvalues.*

Study of Table 6.5 clearly shows that the roundness factors are of greater importance compared to sphericity factors in three environments. However, in dune environment SSN factor is more important which may be due to extreme selective sorting processes on the basis of sphericity in wind transport. Wind selects grains of higher sphericity to be easily transported as bed and saltation load, thereby nonnormality in the sphericity distribution results. In river environment RNN factor is most important, which is interpreted as abrasion processes did not operate with sufficient intensity or for sufficient amount of time to produce normal distribution of roundness in river samples. The RM-RSK factor is generally very important in all three environments but assumes greater importance in the beach environment. We may interpret that abrasion processes affecting roundness are important in all three environments whereas selective sorting on the basis of sphericity decrease in importance as dune-beach-river. The factors SM and SSD have less importance, which is interpreted as lack of sphericity changes in the fine sand ranges. The RSD factor is of medium importance in all environments, which indicates that roundness changes (due to abrasion) do occur in fine sand size ranges. More details of interpretation are included in a paper by Patro and Sahu (1974).

Table 6.6 Mean values of different variables in each environment (on basis of $\log(\Psi/(1-\Psi))$ and $\log P$)

Environment → Variable ↓	River	Beach	Dune
1. Sphericity Mean (SM).	0.2291	0.2370	0.2529
2. Sph. Stand. Dev. (SSD)	0.1768	0.1686	0.1723
3. Sph. Skewness (SSK)	–0.0276	–0.0601	–0.0775
4. Sph. Kurtosis (SK) -0.0455	–0.1477	–0.1773	–0.1773
5. Roundness Mean (RM)	0.6582	0.6886	0.6181
6. Roundness Stand. Dev (RSD)	0.1507	0.1516	0.1496
7. Roundness Skewness (RSK)	–0.1077	–0.1536	–0.1124
8. Roundness Kurtosis (RK)	–0.0818	0.1822	–0.1287

6.4 Example : (*Chi-Square Test for Goodness of Fit*)

We reinterpret an example given in Griffiths (1967, p345-348) where χ^2-test is used to fit an unimodal phi-normal (log-normal) size distribution from 450 quartz short axes (c_ϕ) data of Homewood quartzite. The *c*-axes data is grouped into phi-size intervals, so that a Normal distribution is to be fitted with parameters $\mu = 1.4278$ and $\sigma^2 = 0.3117$ ($\sigma = .5583$). The results are shown in the table below (but regrouping was done to keep expected frequencies to 1 or more for each group).

Phi-size mineral	Normal Exp. Prob	Normal Exp. Frequency (E) (NEP)	Observed frequency(O) (NEP) × 450	Obs χ^2 (O – E²)/E	Inference on Transportation and Deposition
<– .5	.00034	0.14	6	5.67	Bed-load component
–.5-0.0	.0049	2.14			
0.0 – 0.5	.0433	19.49	12	2.88	
0.5 – 1.0	.1721	77.45	59	4.40	
1.0 – 1.5	.3311	149.00	191	11.84	Saltation component
1.5 – 2.0	.2968	133.56	117	2.05	
2.0 – 2.5	.1241	55.86	55	0.01	
2.5 – 3.0	.0250	11.25	7	1.61	
3.0 – 3.5	.0023	1.04	3	3.35	Suspension component
>3.5	.0001	0.05			
Total	1.000	450.00	450	$\chi^2 = 31.81$	

The degrees of freedom for χ^2 with 8-1-2 (parameters) = 5 d.f. and $\chi^{0.5}_{5,\,0.01} = 20.517$. Thus, the observed chi-square is much greater than 20.517 and does not fit a unimodal phi-normal distribution with parameters $\mu = 1.4278$ and $\sigma^2 = 0.3117$. Careful study of individual chi-square values in column 5 shows that it has two minimums in row (2) and row (6) with a maximum occuring at row (4) (value 11.84). This indicates that Homewood quartzite has a tri(phi)modal size distribution with phi-normal components such as deposits from bed-load, saltation and suspension loads.

6.5 Exercises

1. The correlation matrix for number of random variables $p = 3$ and sample size $N = 50$ is given

by :
$$\begin{bmatrix} 1.00 & 0.50 & 0.30 \\ 0.50 & 1.00 & 0.20 \\ 0.30 & 0.20 & 1.00 \end{bmatrix}$$

(a) What is the rank of the correlation matrix? Are all the variables linearly correlated ?

(b) Find the eigenvalues of the correlation matrix

(c) Is the third eigenvalue (= 0.49) statistically significant?

(d) What is the probability that $R = I$?

2. The two factor structure of the above matrix is given by ·

$$S = \begin{bmatrix} 1.00 & 0.50 & 0.30 \\ 0.50 & 1.00 & 020 \\ 0.30 & 0.20 & 1.00 \end{bmatrix}$$

(a) Which loadings are statistically nonsignificant?

(b) Indicate and explain the identity of second factor in terms of original variables.

3. Find the varimax rotation of the above factor structure and identify the two factors in terms of the variables.

4. The following are the varimax rotated loadings of three common factors on eight variables. Which variable can be dropped as input and why ?

	Varimax loadings on each variable							
Factor 1	−.7	.1	−.2	.5	.8	−.3	.1	.2
Factor 2	.3	−.1	−.8	.1	−.1	.7	.2	.2
Factor 3	−.2	−.2	.2	.7	.2	.3	-.1	-.7

Answer :

Variables 2 and 7 may be dropped as each of their loadings with common factors is nonsignificant.

Factor 1 is bipolar and loaded on variables 5, 4 and 1.

Factor 2 is bipolar and loaded on variables 3 and 6.

Factor 3 is bipolar and loaded on variables 4 and 8.

So, variables 1,3,4,5,6 and 8 can not be dropped as inputs.

5. The following gives the varimax rotated loading matrix for nine chemical elements in the 32 soils in the Mumbai metropolis area having industrial, vehicular and sewerage pollution. Interpret each varimax-rotated factor in terms of natural weathering of source rocks (basalts) and polluting factors.

Variable → Factor ↓	Cd	Co	Cr	Cu	Fe	Mn	Ni	Pb	Zn	% of variance
1.	.793	.839	.777	.003	.276	.796	.881	.437	.144	40.37
2.	.002	.126	.008	.903	.750	.220	.089	.189	−.121	17.78
3.	.146	.432	−.462	.002	−.127	.165	−.004	−.023	.927	14.71
Communality	.65	.91	.82	.81	.65	.71	.78	.227	.89	72.86

Answer :

F1(unipolar) is loaded on natural weathering of basalts (Co-Cr-Ni) and on anthropogenic factors of industries, vehicles, quarrying (Cd – Mn – Pb). F2(unipolar) is loaded on natural weathering of basalts (Cu-Fe). F3(bipolar) is loaded on Zn due to sewerage and vehicular exhaust pollution.

6. The varimax rotated factor loadings matrix for chemical data of soils/ laterites of Goa area having iron ore, bauxite, Mn mines is given below. Which elements may be eliminated for analysis and why ? Identify and explain each rotated factor (*N* = 50).

Variables	F1	F2	F3	F4	Communality
Na	–.142	–.161	.706	.038	.546
Mg	.898	.260	–.050	–.051	.878
Al	.906	–.178	.087	-.053	.863
Si	.319	.887	–.011	–.008	.888
P	.285	–.461	–.299	.182	.416
K	.900	.127	–.027	–.108	.838
Ca	.653	–.018	–.319	.086	.536
Ti	.721	.449	–.031	.178	.754
Mn	–.301	–.637	–.166	–.326	.629
Fe	–.501	–.804	–.086	.009	.904
Ni	.159	–.366	–.619	–.148	.564
Co	.140	–.389	.120	–.650	.607
Pb	.162	–.828	.121	–.134	.744
Cr	-.427	–.050	–.587	.231	.583
Zn	.051	–.068	.081	.884	.794
%Variance Explained	27.9	22.4	10.01	9.96	70.35

Answer :

Based on communalities, we may consider dropping variables Na, P, Ca from factor analysis. Rotated factor F1 is positively loaded on Si but negatively loaded on Mn, Fe, and Pb. This may be explained by weathering and mining of rocks such as iron ores, Mn-ores, which also contain some sulfides of iron, lead etc. Rotated F3 is bipolar with positive loadings on Na and negative loadings on Ni and Cr. This may be due to seawater encroachment onto agricultural lands which are low-lying and residual fertilizers in these agricultural fields. Rotated F4 is bipolar with negative loadings on Pb and positive loadings on Zn. This is due to vehicular exhaust pollution.

References and Further Reading

1. Bartlett, M.S., Tests for significant factor analysis, British Journal of Psychology (Statistics Section) 3: 77-85
2. Griffiths, J.C., A genetic model for the interpretation of petrology of detrital sediments, Journal of Geology, 74: 658-672
3. Kaiser, H.F., 1958, The Varimax Criteria for Analytic rotation in factor analysis, Psychometrika, 23:187-200
4. Patro, B.C., 1970, A study of size,shape and roundness data of clastic quartz grains , PhD thesis , Sambalpur University, 210p
5. Patro, B.C. and Sahu, B.K., 1974, Factor Analysis of sphericity and roundness data of clastic quartz grains, environmental significance, Sed. Geology, 11:59-74
6. Rao, C.R., 1965, Linear Statistical Inference and its applications, NewYork, Wiley
7. Ratha,D.S. and Sahu,B.K., 1994, Source identification of geochemical variables in sediments of two estuaries of Bombay, India through statistical analysis.
   In: A. G. Fabbri and J.J.Royer (ed.), 3rd CODATA Conf. Sci de la Terra, Nancy, France, 32:383-404
8. Sahu,B.K., 1973, Factor Model applied to sphericity and roundness data of quartz grains, Bulletin of Indian Geological Association, Chandigarh, 6:3-11
9. Sahu,B.K. and Patro, 1970, Treatment of sphericity and roundness data of clastic quartz grains, Sedimentology, 14:51-66
10. Sampson,R.J., 1968, R-mode computer analysis program, Computer Contribution Series, Kansas Geological Survey, 20:13-37

7 Multiple Regression and Canonical Correlations

7.1 Introduction

Multiple correlations include polynomial regressions and these are linear on regression (correlations) parameters (slopes). The squared multiple correlations R^2 indicate the amount of criteria variance explained by multiple (polynomial) regression /correlation and the residual criterion variance is considered to be error variance. We should extract as much explained variance, which should be statistically significant whereas the error variance shall be minimized and statistically non-significant. Since the criterion is linearly related to several predictor variables, we need partial correlation of criterion with each prediction variable keeping other predictors mathematically constant, so that the importance of that predictor can be correctly assessed. Mathematically, this is an univariate statistical procedure as predictors are not random variables, and will not be treated in this chapter.

For canonical correlation, two or more sets of variables one sets criterion and other sets predictors are to be correlated. The eigenvalues define the direction of maximum covariance (correlation) between the criterion prediction sets and in contrast, in principal components and factor analysis the eigenvectors are drawn along directions of maximum variance. If the canonical roots (eigenvalues) are statistically significant, the canonical prediction of criterion from the prediction set (s) is meaningful. Hence, only one or two the predictor variables are also considered random vectors. However, canonical correlation analysis involves finding the eigenstructure of a real nonsymmetric matrix such as $R_{22}^{-1} R_{21} R_{11}^{-1} R_{12}$ which is the product of two real symmetric matrices, B^{-1} and A.

where $B = R_{22}$ and $A = R_{21} R_{11}^{-1} R_{12}$. (Fig. 7.1)

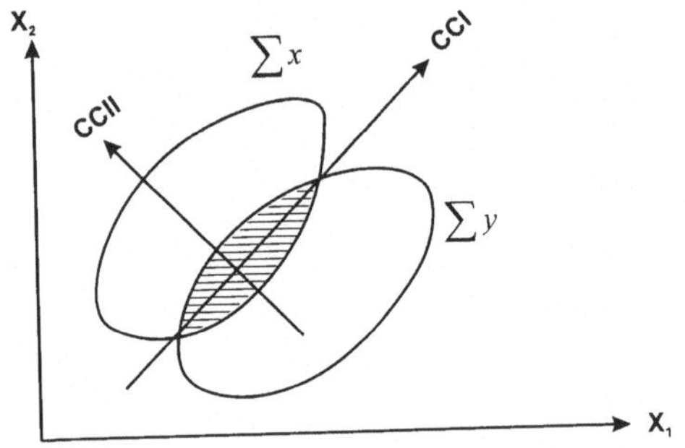

Fig. 7.1 Canonical correlations CC I and CC II based on maximum covariance between vectors

Mathematically we wish to have eigenstructure of real non-symmetric matrix of the form $(A - \lambda_j B)\,\vartheta_j = 0$ or $B^{-1}AV = V\Lambda$. First we solve the eigenstructure of B $(= R_{22})$ as $BU = U\Lambda_1$.

Then $$B = U\Lambda_1 U' = \left(U\Lambda_1^{\frac{1}{2}}\right)\left(\Lambda_1^{\frac{1}{2}} U'\right) = B^{\frac{1}{2}} B^{\frac{1}{2}}$$

We know $$B^{-1} = U\Lambda_1^{-1} U' = \left(U\Lambda_1^{\frac{1}{2}}\right)\left(\Lambda_1^{\frac{1}{2}} U'\right), \text{ so } B^{-\frac{1}{2}} = U\Lambda_1^{\frac{1}{2}}$$

Substituting $B^{-\frac{1}{2}}$ in canonical equation we have $$\left(A - \lambda_j U\Lambda_1^{\frac{1}{2}} U'\right)\vartheta_j = 0$$

or, $$\left(\Lambda_1^{\frac{1}{2}} U' A U \Lambda_1^{-\frac{1}{2}} - \lambda_j I\right)\Lambda_1^{\frac{1}{2}} U' \vartheta_j = 0$$

or $$\left(B^{-\frac{1}{2}} A B^{-\frac{1}{2}} - \lambda_j I\right)\left(B^{\frac{1}{2}} \vartheta_j\right) = 0$$

Solving eigenstructure of real symmetric matrix $B^{-\frac{1}{2}} \wedge B^{\frac{1}{2}}$ gives the eigenvalues of $B^{-1}A$. The resulting eigenvalues $\left(B^{-\frac{1}{2}} V\right)$ are reduced to eigenvectors V of $B^{-1}A$ by

$$V = B^{-\frac{1}{2}} \left(B^{-\frac{1}{2}} V\right)$$

The statistical significance of diagonal elements (canonical eigenvalues) for matrix $B^{-1}A$ can be found as follows.

Proportion explained by $\lambda_j = \lambda_j / \text{tr } \Lambda$ and is retained if it is high, otherwise dropped as error.

Bartlett (1947) defines a test statistics lambda as

$$\Lambda = \prod_{J=1}^{P_2} \left(1 - \lambda_J\right)$$

where is the dimension of second set.

The null hypothesis (H_0) that z_1, is unrelated to z_2 is tested as a χ^2 variable with p_1 p_2 df :

$$\chi^2 = \left[(N - 1) - \frac{1}{2}(p_1 + p_2 + 1)\right] \log_e \Lambda$$

If the null hypothesis is rejected, then effects of λ_1(first canonical root) is removed and rest (p_2 – 1). Canonical roots are tested as $\Lambda' = \prod_{r+1}^{P2} \left(1 - \lambda_j\right)$ and

$\chi^2 = \left[(N - 1) - \frac{1}{2}(p_1 + p_2 + 1)\right] \log_e \Lambda$ with (p_1 – 1) (p_2 – 1) df and so on.

In general, r canonical relations can be removed as $\Lambda' = \prod_{r+1}^{P2} \left(1 - \lambda_j\right)$ and

$$\chi^2_{(p_1 - r)(p_2 - r)} = -\left[(N - 1) - \frac{1}{2}(p_1 + p_2 1)\right] \log_e \Lambda'.$$

Usually any canonical root less than 0.30 may be statistically nonsignificant and hence can be dropped.

A multiple regression equation in standardized variables can be written as $\hat{Z} = b_1 z_1 + b_2 z_2 + \ldots b_{L-1} z_{p-1}$ and coefficients of multiple correlation $R_{P, 1, 2, \ldots P-1}$ (= R) is similar to product-moment correlation coefficient $-1 \le r \le 1$ but ranging $0 \le R \le 1$. R^2 explains the variance of criterion variable and $(1 - R^2)$ is the amount of unexplained regression (correlation) error. Therefore we have the F -test for multiple correlation equation as

$$F_{N-p}^{p-1} = R^2 (n - P)/(1 - R^2)(p - 1)$$

The standard error for z_p is $se_z = \sqrt{1 - R^2}$

The p × p correlation matrix for all z_p variables is denoted as R which is positioned as R_{11} with (p – 1) × (p + 1) correlation matrix of predictors variables and z_p (criterion) has a variance = 1.

Thus, $$R = \left[\begin{array}{cc} R_{11} & R_{12} \\ \hline R_{21} & 1.0 \end{array}\right]$$

we get the slopes as $b = R_{11}^{-1} R_{12}$. However high values of b_i are not necessarily confirming that variable z_i is strongly predicting criterion z_p as other variables are simultaneously present to confound our judgment.

In polynomial regression the predictors are powers of a random variables x as $z_1 = x$ (linear), $z_2 = x^2$ (hyperbolic), $z_3 = x^3$ (sinusoidal) etc. We should not make polynomial regression for powers four and more, unless there is strong mathematically theory behind the higher power polynomial regressions. Higher degree polynomial fit should have significant F-test regression variance explained and includes all lower order polynomial terms (which may or may not be individually significant by F-tests).

In order to study multiple partial correlations the input vector z of order p can be partitioned into three subsets as z_1 (control variables whose effects as criterion z_2 is to be removed), z_2 (criterion) and z_3 (independent predictors).

Then correlation matrix

$$R = \begin{bmatrix} R_{11} & R_{12} & R_{13} \\ R_{21} & R_{22} & R_{23} \\ R_{31} & R_{32} & R_{33} \end{bmatrix}$$

We obtain standardized regression with $B = R_{11}^{-1} R_{12}$. Next we compute the covariance matrix of $\hat{Z}_2$ as

$$\hat{R}_{22} = R_{21}\ B = R_{21}\ R_{11}^{-1}\ R_{12}$$

The residual matrix

$$\tilde{R}_{22} = R_{22}\ \hat{R}_{22} = R_{22} - R_{21} R_{11}^{-1} R_{12}$$

$$R_{2.1} = \{r_{2.1\ jk}\} = \hat{r}_{22jk} / \sqrt{\hat{r}_{22}\ \hat{r}_{22\ kk}}$$

Then we form multiple partial correlation matrix as

$$R_{2.1} = \{r_{2.1\ jk}\} = \bar{r}_{22\ jk} / \sqrt{\bar{r}_{22\ jj}\ r_{22kk}}$$

If z_3 is a vector but z_1 and z_2 are scalar then

$$\bar{r}_{22} = 1 - r_{21}^2 \quad \text{and} \quad \bar{r}_{23j} = \bar{r}_{23j}/\sqrt{1 - r_{21}^2}$$

so that $r_{21.3\ j} = \bar{r}_{23\ j}/\sqrt{1 - r_{21}^2}$ which is a well-known result.

Here $\bar{R} = R_{23} - R_{21}\ R_{11}^{-1}\ R_{13}\ \bar{r}_{23\ jk}$ are jk^{th} covariances in

$$\bar{R}_{23} = R_{23} - R_{21}\ R_{11}^{-1}\ R_{13}$$

We can also compute $R_{21.3j} = \{r_{2.1,3}\} = \bar{r}_{23\ jk}/\sqrt{\bar{r}_{22\ jj}}$

The output of partial correlation can be arranged as

$$R = \left[\begin{array}{cc} R_{21} & R_{21.3} \\ \hline R_{3.2.1} & R_{33} \end{array}\right]$$

Partial correlations are necessary to find relation between two variables by keeping other concomitant predictor variables mathematically constant. This gives the true importance of the predictor variable for prediction of criterion variable.

7.2 Case Study

Evaluation of pollutants in soil and sediments of Bombay region, using canonical correlation.

Sediments in aquatic systems sorb many trace elements, pollutants, nutrients through ion exchange, adsorption, and/or desorption depending on the physicochemical conditions and the species of the pollutants or trace elements. Sorption is inversely related to the particle size of the sediments. Size parameters viz. mean (M), standard deviation (SD), skewness (Sk) and kurtosis (K) control the concentration of trace elements in sediments. Further concentration is directly proportional to amount and type of clay fraction. The effect of size parameters on the concentration of trace elements was studied by multivariate canonical correlation analysis. The choice of such an analysis is that in an eco-sedimentological system, some variables will function as predictors while other variables will function as responses (dependent upon the predictors).

Table 7.1 Trace element data for sediments (average concentration in ppm)

Sample*	Cd	Co	Cr	Cu	Fe	Mn	Ni	Pb	Zn
U1	2.63	48.67	59.67	84.00	53,486.67	380.67	97.00	37.00	58.33
U2	2.90	45.33	78.67	101.00	55,896.67	368.67	101.33	100.67	128.33
U3	3.87	48.00	60.33	87.33	53,580.00	428.00	90.67	50.00	141.00
U4	3.63	45.00	57.00	81.67	59,776.67	365.00	84.33	47.00	135.33
U5	3.63	42.00	65.67	76.67	53,860.00	387.33	89.67	47.33	121.67
U6	3.23	45.33	100.6	96.00	58,620.00	481.67	88.67	46.00	136.37
U7	3.80	43.00	76.00	101.00	58,133.33	472.67	93.00	43.67	142.00
U8	4.10	48.00	81.00	126.33	61,026.67	480.00	95.33	50.00	188.67
U9	2.83	41.00	74.67	117.00	59,616.67	472.33	110.00	54.33	178.00
U10	2.77	42.67	88.00	132.00	57,826.67	493.33	108.00	45.67	200.67
U11	2.67	46.67	89.33	123.00	54,886.67	495.67	106.00	52.00	194.67
BG1	2.00	59.00	34.67	62.67	25,636.67	24.33	59.67	54.67	65.67
BG2	2.47	55.33	53.00	79.67	29,390.00	46.33	85.00	66.67	88.33

U: Ulhas river samples; BG1, BG2 : background samples

Table 7.2 Physical size parameters

Sample	Mean	Standard deviation	Skewness	Kurtosis
U1	6.866	2.983	1.847	0.586
U2	7.333	2.813	1.669	0.552
U3	7.366	2.405	1.449	0.554
U4	7.660	2.098	1.753	0.640
U5	7.550	3.002	1.746	0.621
U6	7.662	3.284	1.513	0.558
U7	8.021	3.228	1.558	0.544
U8	7.712	2.341	1.446	0.555
U9	7.682	2.271	1.338	0.562
U10	7.662	3.283	1.512	0.558
U11	7.332	2.813	1.667	0.549
BG1	8.228	3.441	1.562	0.512
BG2	8.312	3.432	1.558	0.556

Three sets of samples, two postmonsoon and one premonsoon, were taken for the purpose of studying the sediments. The sediments were collected from Ulhas river (located in a thickly industrialized belt) and trace elements (viz. cadmium, cobalt, chromium, copper, iron, nickel, manganese, zinc, lead) were determined by atomic absorption spectrophotometry after digestion of the powdered sample (200 mesh) in a 1:3 concentrated perchloric-nitric acid mixture. Here, only the average concentration values were taken into consideration. Size analysis was carried out by hydrometer and sieve analysis. Size parameters were calculated from the cumulative size distribution curves using graphic distribution formula of Folk and Ward. Prior to the canonical

9 789352 300327

Printed at Repro India Ltd.